MW00964265

THE POLITICS OF DEVELOPMENT COOPERATION

The Politics of Development Cooperation interrogates the politics of inter-organizational development cooperation, examining issues of power, autonomy, and dependence. Focusing on Kenya and in particular on Maendeleo Ya Wanawake (MYWO), the largest national women's organization, and its partners in its relational environment, this book probes the relationships between foreign donors, grassroots development organizations and governments.

Aubrey examines whether it is possible for the North and the developing world to be engaged in genuine development partnerships, the influence resource contributions, financial and technical, have on agenda formulation and compromises, and whether organizations such as MYWO are truly NGOs, as they claim to be, or whether they remain an extension of the state exploited by patriarchal party politics.

Gender is central to the analysis of this book, with issues reflecting and reintroducing the politics of unequal resources in development cooperative partnerships. Differences in status among women are also systematically examined because the politics of development affect elite and grassroots women differently.

Lisa Aubrey is an Assistant Professor in the Department of Political Science and the African Studies Program at Ohio University.

ROUTLEDGE STUDIES IN DEVELOPMENT AND SOCIETY

1 SEARCHING FOR SECURITY
Women's Responses to Economic Transformations
Edited by Isa Baud and Ines Smyth

2 THE LIFE REGION
Edited by Per Råberg

3 DAMS AS AID
Anne Usher

4 POLITICS OF DEVELOPMENT COOPERATION
Lisa Aubrey

THE POLITICS OF DEVELOPMENT COOPERATION

NGOs, Gender and Partnership in Kenya

Lisa Aubrey

London and New York

First published 1997
by Routledge
11 New Fetter Lane, London EC4P 4EE

Simultaneously published in the USA and Canada
by Routledge
29 West 35th Street, New York, NY 10001

Typeset in Garamond by Routledge

Printed and bound in Great Britain by
TJ International, Padstow, Cornwall

British Library Cataloguing in Publication Data
A catalogue record for this book is available from the British Library

Library of Congress Cataloguing in Publication Data
Aubrey, Lisa Marie, 1961–
The politics of development co-operation : NGOs, gender and partnership in Kenya / Lisa Aubrey.
(Routledge studies in development & society)
Includes bibliographical references and index.
1. Economic development–International cooperation–Case studies. 2. Kenya–Economic policy. 3. Women in development–Kenya. 4. Non-governmental organizations–Kenya. I. Title. II. Series: Routledge Studies in development and society.
HD75.A923 1997
338.9–dc21 97–14343
CIP

ISBN 0–415–15185–6

To my daughter Kaari and my parents Luke and Shirley Aubrey, who taught me that partnership in a spirit of *genuine* cooperation is key.

CONTENTS

List of illustrations viii
Preface ix
Acknowledgements xi
Chronology xv
Acronyms xvii

1 INTRODUCTION 1

2 BRIDGING LITERATURE GAPS: FRAMING THE PROBLEM OF THE POLITICS OF DEVELOPMENT COOPERATION 9

3 THE EVOLUTION OF MYWO FROM 1952 TO 1992 45

4 A CHANGING RESEARCH METHODOLOGY AMID POLITICAL VOLATILITY, ENVIRONMENTAL UNCERTAINTY AND A CULTURE OF FEAR AND SILENCE 89

5 RESEARCH FINDINGS: A BARRAGE OF CONTRADICTIONS 108

6 THE WEB OF DECEIT: GRASSROOTS DEVELOPMENT CAUGHT? 143

APPENDICES
Appendix A.I *Hypotheses* 169
Appendix A.II *Sub-hypotheses* 170
Appendix A.III *Key variables and their definitions* 172
Appendix B *Survey instrument: list of open-ended questions* 175
Appendix C *Foreign donors* 179
Appendix D *Kenyan government ministries* 183

Notes 185
Bibliography 214
Index 229

LIST OF ILLUSTRATIONS

Map Kenya: administrative boundaries xiv

FIGURES

2.1 An organization's relations to its environments 18
2.2 Interorganizational system 42
2.3 Application of general systems theory to this study 42
2.4 Transnational network linkages 43
A.1 Presentation of hypotheses proposed 172

TABLES

6.1 Foreign donors funding MYWO programs/projects 150

PREFACE

The seed for this book was planted and nurtured by many different sources. One source—which is the strongest—is the refusal to believe that Africa has little chance of developing, despite what the quantifiable data tell us. A second source is a deep-seated hope that perhaps *just* the right approach to development, theoretical and practical, will eventually be developed and tried, emanating from within the continent. A third source was my desire, in this endeavor (and in all other endeavors in my life) to engage a topic that had not only scholarly merit, but also just as much practical importance for people's lives and for my life.

The development literature of the late 1980s was the fertile soil in which I planted this seed. Its foci on people's participatory movements, gender and grassroots awakening, and indigenization was exciting and new, and, I believed, offered some promise for Africa's development. Further capturing my interest, because of its potential for development, was literature I came across at Indiana University by and about Maendeleo Ya Wanawake (MYWO). The literature expressed MYWO's commitment to women's progress and grassroots development in Kenya. MYWO seemed to fit the model of development from below, formulating an indigenous development agenda from the grassroots. I envisioned MYWO holding the banner of gender-sensitive indigenization: women leading development in Kenya from the grassroots. I was disappointed when I went to the field.

A much-needed resource for nurturing the seed of this project for me was a relational context within which to place MYWO. Why was studying MYWO in Kenya important to the larger human population? Why and how do the activities of MYWO affect me? As members of the human family we are all interconnected by international and transnational relations; hence looking at MYWO in the partnerships within which it interacts seemed to be fitting for studying the nature of our interdependence as we move to the next millennium. None of us exists alone: not nations, not states, not organizations nor individuals. In fact, global cooperation may be the only way we humanly develop.

Securing research permission from the Kenyan government to carry out this study in the field was a process that took longer than two years. The research experience and results of this study, as well as this book, have made the wait worthwhile. My only disappointment is that, in reality, MYWO was not a grassroots organization, leading gender-sensitive development and working toward women's progress in Kenya. In fact, MYWO proved to be just the opposite.

I returned to the field in the summer of 1996 to discuss my findings and conclusions before the publication of this book. It was an important methodological step, to me, to get interviewees' feedback and critique. The feedback was overwhelmingly positive and supportive of the analyses and conclusions that I reached, particularly among the women in rural areas. MYWO national elected officials were in transition and most were not available to meet with me. Upon my return, I received a letter from them inviting me to take a look at the changes in the organization since the installation of a new leadership. I have begun to do that.

Returning to the field granted me the assurance that the views of interviewees are adequately and fairly represented throughout this book. Many of the interviewees, however, reiterated that they would like to remain anonymous. I have done my best to honor their requests and thank them for ensuring that the seed that was planted has blossomed to fruition.

ACKNOWLEDGEMENTS

There are many to whom I am indebted, for it is their encouragement and assistance which has taken me from the initial stages of this project through its completion. Without each and every one of these persons, the culmination of this project would not have been possible. Dr Chadwick Alger has been a pillar of support and an intellectual mentor for me since our paths merged in 1989. His work on non-governmental organizations has given me remarkable insights into development cooperation. Dr William Liddle exposed me to development literature as a graduate student. He challenged me and also encouraged me to send this manuscript out for publication. He also suggested the title of the book. To Dr Isaac Mowoe, I express heartfelt gratitude for mentoring, for career and life advisement, and for friendship. Our work together since the anti-apartheid movement in 1985 has served as an arena in which I have been able to learn from his intelligence, diplomacy, sensitivity and fairness.

Gratitude is also due friends and relatives who are too many to name here, but who unselfishly offered their assistance and encouraged me to do field research when I began to ask questions about grassroots organizations in Africa. Among those I must especially thank are Nick Saunders and my sister Merinda.

I am appreciative of the assistance of the Social Science Research Council, the Institute for the Study of World Politics, and the True Friends of LaPointe, Louisiana Benevolent Association for making it possible for me to spend one year in the field. I am also appreciative of the assistance I received from the National Endowment for the Humanities for making it possible for me to have sufficient time to write this study.

This book has taken years to evolve. Before taking a position at Ohio University where I currently teach, I was attached to both the University of Southwestern Louisiana and Southern University, Baton Rouge. At the University of Southwestern Louisiana, I would like to express my gratitude to Dr Janet Frantz for providing me with "a space of my own" so that I could write, and for providing critical feedback on all of my chapters, as well as for listening to my questions and providing suggestions as I tried to analyze my field experience. I am also grateful to Debbie Olivier, also of the University of Southwestern Louisiana, for her technical assistance. To my colleagues at Southern University, Baton Rouge, I am thankful for their encouragement, particularly that of Dr Kingsley Esedo.

At Ohio University, I would like to thank the chair of the Department of Political Science, Dr David Williams, and other colleagues for their encouragement and support, as well as the College of Arts and Sciences who, with my department, granted me a course reduction to work on the completion of the manuscript. I would also like to thank the secretarial staff for their assistance. I am further indebted to the African Studies Program and its Director, Dr Steve Howard, for the opportunity to return to the field to do the follow-up methodological work. I would also like to thank him for his encouragement "to get the book out." I would like to thank the Cartography Center, especially Steve Dishong, for providing the map for this book.

My colleagues in the Bentley Hall basement also provided intellectual and moral support. I would like to thank Mary Anne Reeves, Dr Sholeh Quinn, and especially Dr Katherine Jellison. I first shared the news of Routledge's interest in my manuscript with Dr Jellison. Since that time, she has walked me through the necessary stages for the completion of the manuscript, encouraging me along the way. For that I am grateful.

There are also many others who nurtured my intellectual curiosity and my interest in Africa. This preparation was the foundation for my writing this book. To Dr Jewel Prestage who first defined politics for me; to Dr Gloria Braxton who first taught me about African politics and who encouraged me to take my first trip to Africa through Operation Crossroads in 1982; and to Dr Yousef Danesh who captured my interests in international relations, transnational relations, and comparative politics as an undergraduate, I am grateful. I would also like to thank Dr Patrick O'Meara and the African Studies Program faculty at Indiana University for their welcoming me into their program as a CIC Traveling Scholar. It was there that I was exposed to the scholarship on politics and development in Africa. I am grateful.

There are also other individuals who engaged my intellectual curiosity and gave of themselves to assist me in this endeavor, and for that I am grateful. They include Dr Ruth Nasimiyu, Michael and Mary Carson, Dr Ibrahim Abdullah, Dr Dickson Eyoh, Dr Yolanda Comedy, Jay Tettenhorst, Enid Fisher, Cynthia Fue, Grace Jennings, Kenyatta Albeny and Andre Simbine. There are others from Kenya whom I am not able to mention here.

There is one colleague to whom I am especially indebted, Dr Rae Ferguson. Our paths met when we were graduate students together at Indiana University. Dr Ferguson, whose work is also on women's groups, albeit in the United States, has encouraged and challenged me from the beginning of this project to the end. She also took on the tremendous and time-consuming responsibility for all editing and typesetting before this book went out to the publisher. I am forever grateful.

I am also very grateful to several of my students who assisted me in the completion of this manuscript, especially Cheryl Cooper and one special other. They were always willing to do whatever was needed, at whatever time it was needed, to make the publication of this book a reality. Husher Harris came on board to assist at a late date but his contribution was indeed commendable. I would like to express my sincerest gratitude.

None of this work would have been possible if I did not have caring childcare providers to help with taking care of my daughter Kaari so that I could concentrate on the completion of the book. I would like to thank my parents, my aunt Monita, my brother Jimmy, my sister-in-law Liz, my cousin Sharon Leonard and friend Tama Hamilton-Wray. I would especially like to thank my sister Tina who took care of Kaari for the summer of 1996, allowing me to return to Kenya. I would also like to express my sincerest gratitude to Jennifer Ridha who took great care of Kaari during the last couple of weeks of writing and editing this book. Kaari will miss her.

To the many women and men in Kenya who gave of their time to be interviewed, I am eternally grateful.

To my editors Valerie Rose, Sarah Lloyd and Matthew Smith at Routledge, I am very thankful for your assistance.

Producing this book, from inception to delivery, has truly been a cooperative process. Many advisors, colleagues, students, friends and family have given direction, advice, time and energy. There is no way to thank them sufficiently for what they have contributed to this project.

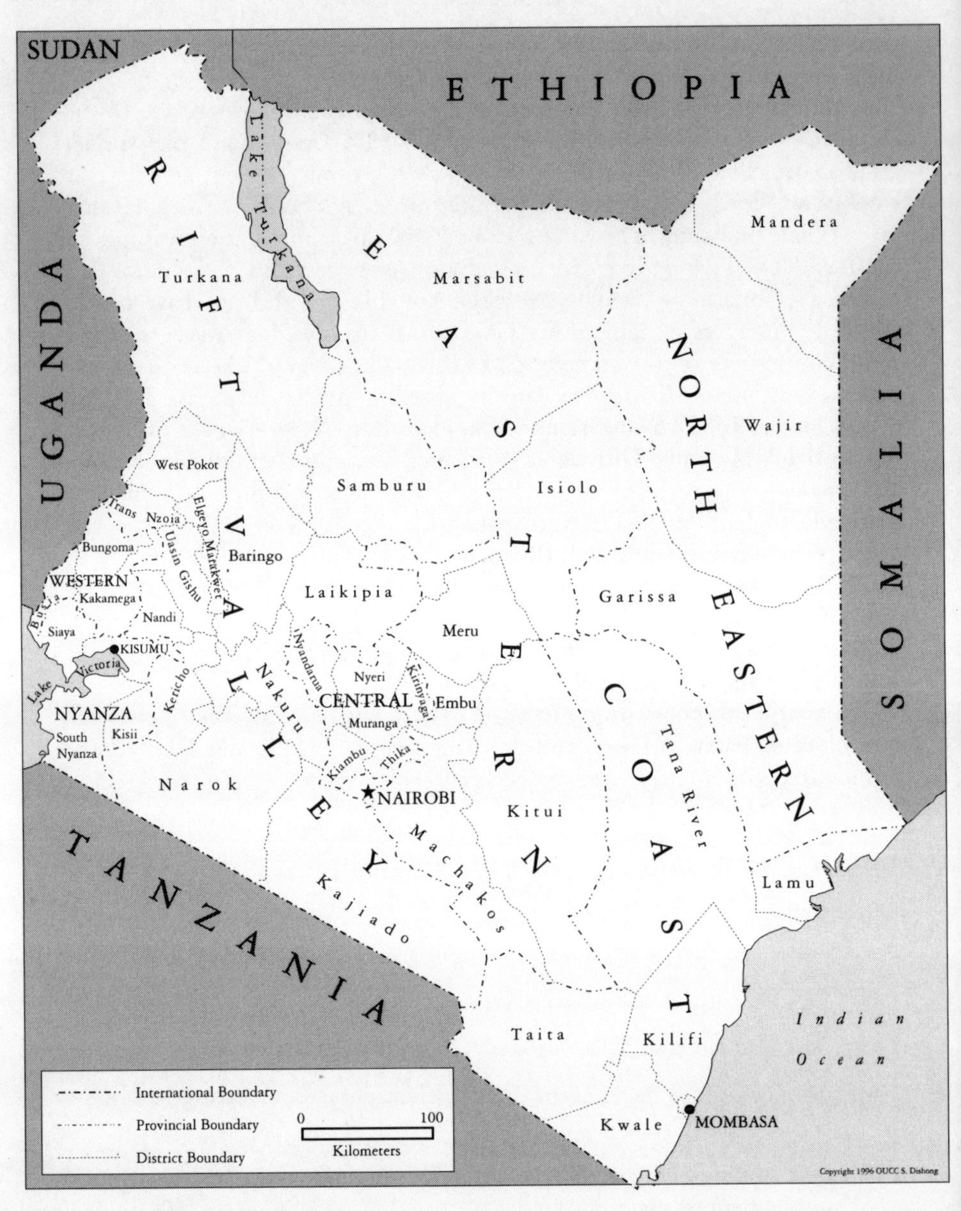

Map Kenya: administrative boundaries

CHRONOLOGY

1952 Maendeleo Ya Wanawake (MYWO) founded by wives and relatives of British Colonial Officers.

1961 Kenyan women take over the leadership of MYWO.

1963 Kenya gains its independence from Britain and Jomo Kenyatta becomes first president of independent Kenya.

1969 Kenya becomes a *de facto* one-party state, with KANU as the sole ruling party.

1982 Kenya becomes a *de jure* one-party state, with KANU as the sole ruling party.

1978 Daniel T. arap Moi becomes president of Kenya after Kenyatta's death.

1984 Jane Kiano resigns as chair of MYWO after 13 years.

1986 Theresa Shitaka formally dismissed as chair of MYWO and a caretaker committee appointed to run MYWO.

1987 MYWO affiliated to KANU.

1989 MYWO election held. Wilkista Onsando becomes chair.

1991 Section 2(a) of the Kenyan Constitution repealed, reverting Kenya to a multi-party state.

1992 MYWO officially disaffiliated from KANU; Kenya presidential and parliamentary elections held.

1996 MYWO elections held.

ACRONYMS

ACWW	Associated Country Women of the World
CBD	Community Based Distributor
CEDPA	Center for Population and Development Activities
CEO	Chief Executive Officer
COTU	Central Organization of Trade Unions
DANIDA	Danish Agency for International Development
DDC	District Development Committee
DEMO	Democratic Movement
DP	Democratic Party
FORD	Forum for the Restoration of Democracy
GAD	Gender and Development
GBM	Green Belt Movement
GEMA	Gikuyu Embu Meru Association
GNP	Gross National Product
GO	Governmental Organization
GRO	Grassroots Organizations
GTZ	German Agency for Technical Cooperation
IGO	International Governmental Organization
IMF	International Monetary Fund
INGO	International Non-Governmental Organization
IOR	Interorganizational Relations
IPK	Islamic Party of Kenya
JICA	Japanese International Cooperation Agency
KADU	Kenya African Democratic Union
KAF	Konrad Adenauer Foundation
KANU	Kenya African National Union
KBC	Kenya Broadcasting Corporation
KMYWO	KANU Maendeleo Ya Wanawake Organization
KPBWC	Kenya Professional and Business Women's Clubs
KPU	Kenya People's Union
KTMT	Kenya Times Media Trust
LD	Leadership Development

LDC	Lesser Developed Countries
LSK	Law Society of Kenya
MCSS	Ministry of Culture and Social Services
MCH/FP	Maternal Child Health and Family Planning Program
MHAH	Ministry of Home Affairs and National Heritage
MNC	Multinational Corporation
MP	Member of Parliament
MYWO	Maendeleo Ya Wanawake Organization
NCCK	National Christian Council of Kenya
NCPD	National Council on Population and Development
NCWK	National Council of Women of Kenya
NGDO	Non-Governmental Development Organization
NGO	Non-Governmental Organization
NORAD	Norwegian Agency for International Development
OAU	Organization of African Unity
OECD	Organization for Economic Cooperation and Development
OXFAM	Oxford Committee for Famine Relief
PRO	Public Record Office
QNGO	Quasi Non-Governmental Organization
RDM	Resource Dependency Model
SAP	Structural Adjustment Program
SDP	Social Democratic Party
SEP	Special Energy Project–Jiko
TNC	Transnational Corporation
UNICEF	United Nations International Children's Emergency Fund
USAID	United States Agency for International Development
VDO	Voluntary Developmental Organization
WAD	Women and Development
WID	Women in Development

I

INTRODUCTION

Development in Africa remains fundamentally grounded in "who gets what, when, how," how much, under what conditions and at what costs.[1] To believe that development is about restructuring the world so that economies will become more balanced, trade will become more fair, education will be more accessible, women will become more equal to men, and all people will be able to better their conditions economically, politically, socially and environmentally, is to define development, as currently being implemented, from an apolitical and unrealistic stance. To believe that those who have plenty will willingly, of their own volition, share equally what they have with those who have less is romantic idealism.

Claude Ake, in *Democracy and Development in Africa*, argues that, "The problem is not so much that development has failed [in Africa] as that it was never really on the agenda in the first place."[2] Politics, from colonial times to the present, has continued to play a central role in development. In fact, politics permeates the most current of endeavors contrived to "bring" development to the continent via development cooperation in the form of development cooperative partnerships. This study is an interrogation of the politics of development cooperation between the partners in development cooperative endeavors in Africa, namely African non-governmental organizations (NGOs), foreign donors and African governments.

The study questions whether or not this new politics of partnerships offers some departure from the liberal modernist approach to development, which imposes the will of the "modern" on the more "traditional."[3] It does so by probing whether or not "partnerships" offer hope for more genuine and less normative development in Africa—development that will better people's life conditions and chances holistically in ways that they deem necessary and appropriate, without the imposition, subtle or overt, of the will of others. It also questions whether or not this new politics of partnerships challenges world systems to eradicate structural inequalities which make development in Africa a preoccupation of so many in the first place.

A LEGACY OF AFRICA'S FAILED DEVELOPMENT POLICIES: THE CHALLENGE FOR SCHOLARS

The challenge for development in Africa is indeed great. Even though more than a quarter of a century has passed since the end of the direct presence of colonial rule on the vast majority of the African continent, prospects for substantial development remain dismal. On the balance sheet of development endeavors, failures of development initiatives far outweigh successes. To many—development planners, recipients, activists and scholars—this means endless frustrations, and makes the need to find solutions to Africa's development problems even more urgent.

Many have attempted to discern the reasons why Africa has failed to develop. The legacy of colonial underdevelopment, neocolonial exploitation of former colonies, internal mismanagement of development resources, the weak political will of African leaders, insincere and vague commitments to development by former colonial powers and post-colonial leaders, natural disasters, misguided economic policies and political instability have all been cited, either singly or concurrently, among the reasons for Africa's continued development malaise.[4]

Just over one decade ago, in 1984–5, this malaise reached endemic proportions with the Ethiopian famine, which the world witnessed. Over one million people died and many more still suffer. The effects of the famine are clearly tragic indeed, yet this is but one example of Africa's current development condition, and, as some would argue, merely one example of an outcome of Africa's development failures.[5] Other disasters had occurred before the Ethiopian famine; many still occur across the continent today. Though some are less publicized, they still are as gripping and devastating. Moreover, the continual deteriorating economic conditions of the continent make the need to find solutions to Africa's development problems even more eminent.

The scramble for solutions to Africa's development problems has led many scholars to rethinking and reshaping the development research agenda. Besides focusing on discerning reasons for Africa's development or non-development predicament, more and more attention is being placed, particularly within the last decade, on trying to rid Africa of its immense human suffering, pervasive poverty and increasing dependence on Western financial institutions, specifically the International Monetary Fund (IMF) and the World Bank.[6] From one perspective, this requires minimizing development failures and increasing development successes. The trilateral partnerships of foreign donors, African governments and African NGOs represent an attempt by development practitioners to do this. Scholars have concomitantly begun to take a research interest in this interorganizational phenomenon.

AFRICA'S HUMAN CRIES: THE PRACTICAL RATIONALE FOR THIS STUDY

The issues raised in this study are not only scholarly in nature but are also of practical importance. They directly address the current impasse in development theory and resultant perplexities about development issues and contemporary global problems. These issues are timely as well. They require, from scholars, critical observations and analyses as development cooperative partnerships possibly lead us into a new era of development (or grassroots colonialism). For development practitioners and much of the world's poor, it is long past the eleventh hour for relieving human suffering and breaking Africa's cycle of poverty and dependence.

As we approach the twenty-first century and as our world becomes smaller and more intertwined through networks of interdependency, it becomes clear that a continent's, a people's, and perhaps the world's future hinges on finding solutions to Africa's development problems. The magnitude and persistence of these problems impelled me to undertake this study. The hard facts are compelling. The 1992 Gross National Product (GNP) of the continent of Africa—south of the Sahara and home to over 600 million people—was almost the same as the GNP of Belgium, a country with 10 million people. Thirty of the world's 40 poorest countries are in Africa. The external debt of the continent has tripled since 1980, now amounting to US$180 billion, a burden so big no one even imagines it can be repaid. Africa's share of world trade is now only 2 percent. Ghana's GNP per capita is only $450 annually, among the lowest on the continent. Between 1980 and 1990, the vast majority of African countries experienced "negative" per capita growth, with few exceptions.[7]

What this means in terms of human suffering is that Africa is ridden with more suffering than the rest of the world, and has an overwhelming majority of its people entrenched in poverty. Approximately 220 million people, more than 1 out of 3, live in "absolute poverty." Four million children born this year will die before the age of 5. Nearly 33 percent of all children are *severely* malnourished. In the past decade, in Zambia, the proportion of children who were malnourished rose from 5 to 25 percent. Over 10,000 children die every day from preventable diseases. The number of children who have the chance of getting an education is declining. Education is not free. Since 1980, the enrollment of children in primary education has dropped from 79 to 67 percent, and national expenditures per student have declined over 33 percent. Food production has also dropped to 20 percent of 1970 levels, and the downward trend continues. Inflation is soaring across the continent; and mothers often cannot buy bread for their children to eat. In addition, poverty is becoming more and more feminized.[8]

How can these problems be resolved, and how can Africa develop? What will it take for Africa, which has the world's largest reservoir of arable land—almost 2.5 billion acres—to feed herself and move towards greater self-reliance? At the beginning of this decade, Pierre Pradervand projected that grassroots NGOs in "partnerships" with donors offered the promise of providing solutions to these problems.[9] This study weighs Pradervand's projection and ponders whether or not interorganizational partnerships between foreign donors, African governments and African NGOs offer possible solutions to the continent's pernicious development problems. It is my belief that unless we systematically examine the relationship between these interorganizational partners cross-nationally and over time, we will perhaps miss the opportunity to understand, evaluate and facilitate development in Africa.

This examination of development cooperative partnerships focuses on Kenya. The case study methodological approach was used to examine the interorganizational relationships of the largest women's organization in Kenya—Maendeleo Ya Wanawake (MYWO)—which also claims to be an NGO. This study examines the relationship between MYWO and its partners in development—foreign donors and the Kenyan government—who provide both financial and technical assistance to MYWO for the implementation of its development programs and projects. MYWO was selected for this study for several reasons. First, MYWO has had long-standing interorganizational relations with foreign donors and the government of Kenya, having initially received financial and technical assistance from the British colonial government and foreign donors prior to its official inception in 1952. Second, it has had extensive countrywide networks extending between its national headquarters in Nairobi and local villages, thereby including grassroots women (local women from rural areas). Third, scholars have engaged discussions about MYWO perhaps more than any other African national women's organization, as an NGO as well as a women's organization threatened by male domination.[10] MYWO is also rumored to be the oldest and largest national women's organization on the continent. Fourth, most of the literature on MYWO, written in the 1970s by sociologist Audrey Wipper, posed contradictions and challenges to our current knowledge regarding the organization's evolution and interorganizational relations, particularly with the Kenyan government. This study aims to broach those contradictions and challenges. In addition, the 1970s literature raised questions about MYWO's autonomy in its interorganizational relationships that researchers have not yet satisfactorily answered. This study attempts to grapple with this issue of autonomy.[11] Fifth, MYWO maintains that it is an NGO. Hence, it is necessary to investigate the legitimacy of its claims, for even after its official affiliation to the Kenya African National Union (KANU) in 1987, MYWO continued to claim that it was non-governmental. Many institutions and individuals in its relational environment, however, argued that it was not. Their questions resurfaced when, in 1992, MYWO was officially disaffiliated from KANU.

The underlying assumption in this study is that in interorganizational development relationships, resource contributions by foreign donors and the Kenyan government to MYWO result in power over MYWO and influence over its development agenda. Hence, the hypothesis follows: the more resources MYWO accepts from foreign donors and the Kenyan government, the less power and autonomy it will have to implement its own indigenous agenda, thereby decreasing the possibility for MYWO's development success. The corollary to this is that the fewer the resources MYWO accepts from foreign donors and the Kenyan government, the more power and autonomy it will have to implement its own indigenous agenda, thereby increasing the possibility for MYWO's development success. Success is defined as the state of reaching the desired objectives and goals of development projects and programs as initially defined by MYWO, before these goals and objectives are influenced by MYWO's development partners.

The key concepts in this study are power, autonomy and dependence. Dependence is defined as the opposite of autonomy. The analytical framework which I applied in this research is a modified version of International Organizational Relations' Resource Dependency Model (RDM). This model was modified so as to draw in the necessary gender perspective which this problem brings to light.

There are two schools of thought which support these hypotheses.[12] The first school advocates the indigenization of development initiatives on the African continent, arguing that indigenization will lead to successful development and eventual self-reliance. In this school, NGOs represent the vessel of indigenization and the hope for development into the twenty-first century. This school highlights the development potential of women, as women are the overwhelming majority of NGO members. Its proponents believe that it is possible for the trilateral partnership of foreign donors, African governments and African NGOs to bring forth development of the continent. This school is perhaps blinded by romanticism.

There is a second school of thought which is much more pessimistic. It suggests that in development cooperative partnerships African NGOs may be vulnerable to Northern NGOs, which this school alleges are but foreign creations to be used and abused so that Africa will remain colonized, not through its leaders as under top-down neocolonialism but this time, in the 1990s, from the bottom up. This school sees Northern NGOs and foreign donors as the new faces of imperialism, engaged in a new form of colonialism of Africa—grassroots colonialism. There are also some in this school who argue that Southern NGOs are also creations of African governments as their means to solidify their political patronage bases. This school does not believe that development is possible through the trilateral development partnerships of foreign donors, African governments and African NGOs. It is my contention that, without thoroughly examining these partnerships, this school is perhaps too suspicious and too premature in its criticisms. It may dismiss *the* answer to Africa's development woes.

With both of these schools, there are obvious conceptual and scientific caveats. My argument here is that a sufficient amount of research examining the systemic trilateral relationship of foreign donors, African governments and African NGOs has yet to be conducted, over time and cross-nationally, and this makes the position of both schools of thought towards interorganizational development partnerships tentative. More research is necessary before either school will be able to make definitive statements about the nature of the interaction between *all* partners in development, or to assess their interorganizational effect on development initiatives in Africa with greater certainty. More fundamentally, as an academic community we are still sorting out the NGO conceptual muddle, and trying to understand and define what NGOs really are and what NGOs really do. Our attempts to address these questions and problems have been noteworthy, however, for they have contributed in a general way to a growing pool of literature which is relevant to development interorganizational relations. The caveat here is that much of this literature, although relevant in context, remains tangential to central questions posed by this study.[13] What is the nature of the coterminous relationship between foreign donors, African governments and African NGOs in development cooperative partnerships? What impact do foreign donor and national government donor contributions to NGOs have on NGO autonomy and ability to implement an indigenous agenda? What effect does NGO autonomy or dependence have on NGOs' ability to formulate and implement successful development projects and/or programs? It is my contention that these questions must be critically examined through systematic observations of development cooperative partnerships in order to construct a model that accurately describes the foreign donor, African government and African NGO relationship. This research intends to contribute to the construction of this model, and perhaps even lay its rudimentary foundation through numerous and diverse intensive field observations and data collection through interviews, archival research and critical analyses.

CHAPTER PREVIEWS

The remainder of this study is presented in the following manner:

Chapter 2

This chapter, "Bridging literature gaps: framing the problem of the politics of development cooperation," will be a review of the relevant literature. This research problem connects four interrelated bodies of literature to provide a context for this study. These are: interorganizational relations theory; NGOs in Africa; development theory; and women in interorganizational relations. Out of these bodies of literature will evolve the framework of analysis for this study.

Chapter 3

Chapter 3, "The evolution of MYWO from 1952 to 1992," will present the political history of MYWO. It will provide a discussion of its changes over time, and document its interorganizational relations. This chapter will also look at MYWO's cyclical pattern, beginning as an appendage organization to the colonial state in 1952, a condition that applied until 1961, and then returning to the post-colonial state as an appendage in 1987. Despite this return to the state, MYWO's constitution continued to claim that it was an NGO. KANU also maintained that MYWO was an NGO. The impact of MYWO's affiliation to both the ruling party and the state on its interorganizational relations with foreign donors will also be analyzed.

Chapter 4

"A changing research methodology amid political volatility, environmental uncertainty and a culture of fear and silence" discusses the evolution of the research methodology used. From August 1991 to August 1992, during the course of my fieldwork in Kenya, the country underwent perhaps its most significant political and economic changes since independence in 1963. These changes inevitably had drastic effects on the nature of the political climate in which I worked, and hence shaped the context of this study. For example, internal and external forces called for the introduction of multipartyism in Kenya. Foreign donors threatened and eventually tied social, political and economic conditions to aid to Kenya—conditions including an end to human rights violations and a call for new elections for the national executive and Parliament. The Moi government threatened to declare a state of emergency, and Kenyan citizens took their cries for "democracy" to the streets. All of these and other events caused me to alter my research strategy, either because I would not be able to execute many of my original research intentions or because they no longer seemed relevant. One barrier that I faced was that very many of the proposed interviewees would not talk, for legitimate fear for their lives. Thus the initial methods proposed for this study differed from the actual methods used. This chapter contains a discussion of both.

Chapter 5

This chapter, "Research findings: a barrage of contradictions," describes the findings of this study and summarizes the data collected as answers to the research questions posed, as well as the data collected from archives. It does not contain references to many of the persons who consented to interviews. Because of promised anonymity, and for the interviewees' personal and professional safety, I do not include names, organizational affiliations or dates of interviews. Their words, however, best reflect their sentiments at the time, as well as their perceptions of important issues. As such, those observations are included (sometimes verbatim) in the chapter.

Chapter 6

"The web of deceit: grassroots development caught?" discusses the findings of this study in light of the current state of knowledge in this field of inquiry. It will tease out the implications of the study for the growing literature on NGOs, for the theoretical bases of interorganizational relations, for development theory and for feminist theory. This chapter also examines the study's limitations. It makes recommendations for optimizing interorganizational contributions for development in Africa and for further scholarly research.

2

BRIDGING LITERATURE GAPS: FRAMING THE PROBLEM OF THE POLITICS OF DEVELOPMENT COOPERATION

This chapter will review the relevant literature from four distinct areas of research and demonstrate their interconnectedness. They are 1) interorganizational relations theory, 2) NGOs in Africa, 3) development theory, and 4) women in interorganizational relations. Although these bodies of literature at first glance seem disparate, particularly the women's studies research, and are rarely integrated in mainstream literature, they are very much interconnected by this study. Out of the merger of these bodies of literature will evolve the framework of the resource dependency model (RDM) which this study will modify and apply to analyze the politics of development cooperation between Maendeleo Ya Wanawake (MYWO) and its partners in its relational environment—foreign donors and the government of Kenya. As the review of literature will establish the basis for the applicability of a modified RDM, the literature will be discussed first, and then the RDM and its modifications will be presented.

As this study focuses on the politics of development cooperation between MYWO, foreign donors and the Kenyan government, specifically the Kenyan bureaucracy, the literature on interorganizational relations (IOR) provides a starting point for the understanding and analysis of interactions between organizations as they work together towards a particular goal. Development interorganizational cooperation is simply "a process in which organizations pursue their own goals and thus retain autonomy, while at the same time orienting their actions towards a common issue or outcome."[1] Thus, IOR theory offers a basis for beginning to systematically understand the relations between organizations as they cooperate in terms of often synonymously used concepts of linkages, networks and partnerships between organizations, and their subsequent effects. David Leonard argues that understanding these concepts is critical for "Linkages [between organizations] can be extremely powerful . . . [since they] are a central component of international aid."[2] Linkages in general, Charles Mulford and Leon Gordenker and Paul Saunders purport, exist because organizations do not operate alone. Inevitably they must interact with other organizations that comprise their clientele, suppliers, supporters, critics and competitors.[3]

There has not been sufficient research attention by scholars to development via interorganizational cooperation among *all* partners in development simultaneously—that is, among foreign donors, NGOs and African governments. There are, however, some studies and research findings about some of the combinations of partnerships (specifically NGOs and the state) which provide a contextual foundation from which to build this study. This chapter will present this literature and build on it. The contextual information that this literature provides will give us insight into why the merger and linkages between foreign donors, NGOs and African governments have not been studied in totality, and will bridge the gaps where the literature is lacking.

STATE OF IOR THEORY

Although a significant amount has been written in the general area of organizations, relatively little attention has been paid to the relationships *between* organizations, or, as it has come to be defined as a field of inquiry, *inter*-organizational relations (IOR). Researchers from various disciplines—particularly psychology, sociology, political science, economics, business administration and management science—have historically concentrated primarily on studying the *intra*organizational phenomenon, and thus the findings they have amassed focus on either: 1) the behaviors of individuals as members in organizations or 2) the more sociological concepts of structures and functions of organizations, rather than the more political concepts of power, autonomy and dependence.[4] In the instances in which the interrelatedness or the relational properties of organizations have been considered, the scope of those studies has been geographically narrow, mainly restricted to the United States in context.

More and more, however, it is becoming necessary to understand the nature and the consequences of *inter*organizational relations globally, for whether we realize it or not all of us—individuals, groups and institutions—interact with and depend on organizations. As students at universities, consumers of health care, members of professional, church or advocacy associations, or even participants in activities that are sponsored by organizations, we are directly and indirectly affected by interorganizational relations in a multiplicity of ways.[5] In addition, solutions to so many of the world's problems such as hunger and destruction of the rain forest depend on interorganizational communication and cooperation. Hence, it becomes imperative that we make a greater effort to understand relationships that develop between organizations. One example of such an effort is L. Julian Efird's 1977 path-breaking study aimed at looking at various international organizations and their interrelatedness in the world food problem area.[6] More recently, Ian Smillie again demonstrated the importance of understanding global and organizational linkages in *The Alms Bazaar*.[7]

The undeveloped state of the theoretical literature on *inter*organizational relations as compared to *intra*organizational relations, as well as its limited focus to the United States, is ironic, especially for the discipline of political science and its subfield, public policy, considering that:

1. Organizations engaged in interorganizational relations are, by their very nature, political. That is, their operations and memberships are political ends, since organizations' objectives include maximizing their control so as not only to realize shared goals, but also to influence, through "lobbying and alliance building," those factors in their environment which cannot be directly controlled;[8]
2. International interactions between and among organizations, units and institutions are the "stuff" international relations as an area of political science is made of;[9]
3. So many basic human activities in the twentieth century are internationalized—communications, technology, science, production of goods, consumption, pricing, development, to name a few;[10]
4. Most, if not all, of the world's problems and their solutions (which I touched upon earlier)—preservation of the environment; eradication of hunger, poverty, homelessness, inequality, AIDS, ending nuclear proliferation and attainment of world peace—transcend national boundaries.

For these very reasons it is surprising how little is known about how we, the world's population, and creators and members of organizations and institutions, are "linked" to each other. Perhaps a greater understanding of how we are in fact "linked" through organizations will bring about an impetus for more systematic studies of IOR. Solutions to our world's problems, only to be reached through individual, organizational and institutional communication and cooperation, depend on it.

IOR SCHOLARSHIP FOCUS—A BIAS TOWARDS DOMESTIC RELATIONS AND AGAINST INTERNATIONAL RELATIONS

One of the reasons that development interorganizational relations between foreign donors, NGOs and African governments have not been a research priority is that there has been a historic bias in academia against international relations and international organizations. For example, Chadwick Alger, in a ten-year review of research on international organizations in the 1960s, made some critical findings which spoke to the state of IOR at that time. He found that of 61 studies, not one focused on relations between international organizations. He did note, however, that "those familiar with the global complex of United Nations activities realize that relations between organizations [were] a matter of increasing concern."[11] Despite this finding, IOR studies that blossomed thereafter were national in scope, not international. Many organizational scholars had broadened their focus from controlling internal activities to managing organizations' external environments, which included other organizations.[12] Concomitantly, from the mid-1960s urban sociologists began redefining urban communities as networks of organizations and relationships with social meaning, as compared to a mere collection of individuals occupying a particular space. Community organizational reference was important, particularly with regard to the delivery of social services, hence the need for interorganizational coordination.[13] That the dynamics of communities could be better understood through IOR remained the state of the field well into the latter years of the 1980s.[14]

Despite the increasing attention to IOR, studies about IOR in the United States far outnumbered studies with an international dimension. For example, research following Alger's ten-year review found that between 1970 and 1974 only one study had been done which examined the interrelatedness of international organizations.[15] A greater number of studies, however, would flourish after 1974, as scholars responded to the recognition of the inevitability of global interdependence and to the remarkable growth of governmental and non-governmental organizations, internationally and nationally respectively. Noting this inevitability, in 1976 a United Nations Conference on Trade and Development spoke of a "world of cooperation" which must be built to reflect the "fundamental interdependence of our destiny."[16] During this same period, by 1978–9, independent international organizations grew to at least 2,700, more than doubling the number that had existed in 1960.[17] In addition, "about 400 U. S. national non-governmental organizations [became] substantially involved in development assistance issues with LDCs" in the South.[18] Specific to the continent of Africa, it is estimated that hundreds of thousands of grassroots African non-governmental organizations existed on the continent during this period. In Kenya in particular, it is posited that between 16,000 and 25,000 organizations sprang up in the 1970s.[19] Subsequently other organizations—governmental and non-governmental donors, primarily from the North—came to their assistance, hence establishing transnational interorganizational linkages in international relations.[20]

ATTEMPTS TO EXPAND IOR FOCUS

Among scholars who attempted to expand IOR's focus beyond domestic organizational relations, particularly because of expanding international relations, Leon Gordenker and Paul Saunders are notable. Their criticism of the academic neglect of the international dimension of interorganizational relations is especially important. They argue that international relations utilizes peculiar forms and processes which the literature on organizations does not discuss, and these differ from those normally conceived about organizations.[21] Hence, in an effort to capture these peculiarities, Christer Jonsson, who agreed with the Gordenker and Saunders critique, proposed that transnational organizational networks serve as the constructs to guide analysis in interorganizational relations. Jonsson defined transnational organizational networks as "national networks of private and public organizations [that] constitute subsystems, and where intergovernmental as well as non-governmental international organizations [GOs and NGOs] typically participate."[22] In this study, in which transnational networks guide the analysis of interorganizational relations in Kenya, it is important to note that particular attention must be paid to the role of the government and the state in this network, as African development policies since colonial times have been heavily statist.[23]

ORGANIZATIONS AND THEIR RELATIONAL ENVIRONMENT

As we approach the twenty-first century it is commonplace that "all organizations have relationships with other organizations" with which they interface and interact.[24] This interfacing and interaction defines their organizational relational environment and grows out of the very fact that many organizations are not self-sufficient and cannot operate alone. An organization's "environment" as defined by Amos Hawley is "all phenomena that are external to and potentially or actually influence the population under study."[25] Thus, most organizations, regardless of their size, scope or reputations, represent only part of a larger system or suprasystem of organizations, within which they must relate, link or connect with other organizations in order to secure the necessary goods and services to maximize their ability to meet their defined goals. This relational environment has come to be conceived in IOR theory as a "social network."

J.C. Mitchell provides a definition of a social network which will serve as a starting point in conceptualizing linkages, connections, relations between and among organizations in this study. He defines a social network as "a specific set of linkages among a defined set of persons with the additional properties that the characteristics of these linkages as a whole may be used to interpret the social behavior of persons involved."[26] Edward Laumann, Joseph Galaskiewicz and Peter Marsden take the initiative in modifying this definition to give it wider applicability to both individual and organizational behavior. They define a social network as "a set of nodes (e.g. persons, organizations) linked by a set of social relationships (e.g. friendship, transfer of funds, overlapping membership) of a specific type."[27] William Evan, Gordenker and Saunders and Jonsson, as discussed earlier, go further with this definition by bringing in an international dimension to the discussions and understandings of IOR, thus making it more applicable for this research. Their contributions basically expand the traditional conceptualization of social networks and widen the purview to apply transnationally.

Howard Aldrich brings to light the practical utility of conceptualizing interorganizational relations in the form of networks. The basic purpose of networks, he argues, is to allow a researcher to track down all the ties that bind organizations in a given population, "as all organizations [are] linked by a specified type of relation, and [are] constructed by finding the ties between all organizations in a population."[28] Conceptualizations of organizational linkages in this way also appear crucial for determining the *outcome* effectiveness of organizational ties. For instance, Andrew Van de Ven and Diane Ferry state that networks are "the total pattern of interrelationships among a cluster of organizations that are meshed together in a social system to attain collective and self interest goals or to resolve specific problems in a target population."[29] Laumann and Franz Pappi's perspective supports this. They surmise that network analysis "assumes that the ways in which nodes [persons or organizations] are connected to one another, both directly and indirectly, influence the behavior of particular nodes and the system as a whole."[30]

IOR, TRANSNATIONAL NETWORKS, RELATIONAL ENVIRONMENTS AND NGOS

For this study, interorganizational relations theorists' assessment of the present state of IOR, despite its caveats, is encouraging. Scholars agree that although what is known about IOR is fragmented, and despite the scholarship being lacking and biased, there is a sufficient amount of rudimentary knowledge to serve as a foundation to guide basic understanding of IOR and further research.[31] They warn, however, that understanding may be challenging, as "IOR analysis is . . . a complex and potentially confusing enterprise, since various studies have different foci, with less clarity in the level of analyses than would be desired."[32] IOR is a further challenge in that it is "often unclear where organizations end and environment begins."[33]

An assessment of the literature on the tripartite relationship between NGOs, governments and foreign donors indicates that, as in the literature on IOR, there is fragmentation. That is, there is a bias towards the examination of the relationship between NGOs and the state, specifically NGOs and states in the South (an expected spillover in the 1990s decade during which Western academics focused on grassroots empowerment, civil society and democratization of the "developing world"). In spite of the contributions made by the literature on NGOs' relations with the state, there is the problematic exclusion of the examination of the interorganizational linkages with foreign donors. This exclusion is problematic because it leaves a gap in the understanding of the overall systemic development cooperative relationship. Julius Nyang'oro points to the more general exclusion of political economy in current African Studies discourses, of which the exclusion of foreign donors in the politics of development cooperation is part.[34] Nyang'oro's criticism is valid, for in neglecting to deal with issues of political economy, African Studies scholarship is neglecting to interrogate the direct correlative relationship between money (and in this specific case, foreign aid) and power in development.

RELEVANT STUDIES

Recent studies of this decade, which include a series on NGOs and the state in Africa, Asia and Latin America by Kate Wellard and James Copestake, John Farrington and David Lewis, and Anthony Bebbington and Graham Thiele respectively, reflect this bias towards unraveling the relationship between NGOs and the state, while at the same time providing valuable, though partial, insight into systemic interorganizational relations.[35] Other studies which tangentially speak to the tripartite IOR issues include Farrington and Bebbington's work on agricultural development, Brian Smith's work on aid and Stephen Ndegwa's work on civil society.[36] Smith's and Ndegwa's studies prove the most relevant to this study, for in addition to both works' tackling of the relations between NGOs and the state, Smith delves into the politics of private aid and Ndegwa addresses the source of Southern NGO organizational power, which he identifies as foreign donor funding. Both Smith and Ndegwa extend their examinations to the multiple nexuses of IOR and raise critical issues which are inextricably tied to the missing link in IOR's current literature—the foreign donor node on the IOR network.

There are also four studies from the late 1980s which address more directly the suprasystem development relationship of NGOs, foreign donors and African governments simultaneously, and which relate more directly to this study. In addition, these works demonstrate the fluidity, unclarity and diverse foci, as well as some inherent conflicts in transnational linkages and cooperative endeavors. Moreover, they demonstrate the importance of understanding organizations and their relational environment in a transnational IOR context.

The first work is Ernesto Garilao's "Indigenous NGOs as strategic institutions: managing the relationship with government and resource agencies." Garilao's major argument is twofold. He focuses on "more stable development NGOs in Asia" and argues that 1) Third World NGOs are more than just conduits for foreign aid, and 2) long-term foreign aid does not lead to NGO survivability; instead it undermines the very existence of the NGO.[37]

Although Garilao's study offers valuable information and insights, it is problematic. First, it does not offer the reader a description of the research methods used to test assumptions/hypotheses and base conclusions. Second, although Garilao proposes to explain the relationship between NGOs and government and NGOs and resource agencies, his study concentrates on the latter relationship—unlike studies aforementioned. The only issue he raises about NGOs and government is that their relationship can shift between collaboration and animosity. He does not expound on this. Third, he flirts with issues of dependence, autonomy and power which lie at the crux of the research problem, but he never addresses them directly. Fourth, he offers prescriptions for NGOs; well-intended as they are, we still do not have a clear understanding of the phenomena with which we are dealing.

The second work is a presentation for the World NGO Symposium, Nagoya Japan Congress, by the United Nations Development Program Assistant Administrator and economist Ryokichi Hirono, entitled "How to promote trilateral cooperation among governments, NGOs and international agencies." Hirono's work is very valuable in that it brings to light relevant issues critical to development cooperation rarely discussed in the literature. Hirono identifies and describes the "attitudes" of mutual hostility and suspicion between the partners in development which, he argues, impede cooperation and hence development. He states that NGOs (which he refers to as voluntary development organizations—VDOs) may fear the power of governments over them, suspecting that governments may try to control or repress them; concomitantly, governments may consider NGOs subversive to national policies and politics. Consequently, governments may resist NGOs' unwelcome participation in politics. Moreover, Hirono states, NGOs may consider foreign donors too bureaucratic, inflexible, establishment-oriented and distant from the poor, while foreign donors may condescendingly consider NGOs amateurish, naive or insignificant. Hirono does not address the relationship between Northern foreign donors and Southern governments.

Hirono does offer prescriptions for increased cooperation between the partners in development. Chief among those prescriptions is the need for NGOs, foreign donors and governments to change their attitudes towards each other. He argues that a simple application of the laws of supply and demand will change negative attitudes. That is, when partners in development recognize "demand" supply will follow. To illustrate, Hirono asserts that there is a demand for partnerships among the trilateral organizations. He argues the following: first, governments need NGOs to expand their development outreach and impact into remote, poor areas; second, foreign donors need NGOs because they are cost-effective and they bring fresh approaches and skills to development; and third, NGOs need foreign donors and government because their assistance allows NGOs to affect national development policies and approaches.[38]

Hirono's presentation makes solutions to a very complex problem seem simple. Although his presentation offers some vital information, "attitudes" of individuals and organizations, as we know, are very difficult to change.

They are embedded in generations of history, culture, tradition and values—all of which may be mutually reinforcing. Moreover, some of the attitudes, such as suspicion of Southern NGOs of the motives of the North or of the sincerity of government, are not unfounded, given the history of their respective relationships. Hirono's supply and demand prescription is also not very realistic. He assumes that all partners are rational actors defining their need, and hence their demand, on the basis of what is good for "development" and for the "whole." We cannot make that assumption, for individual motivation, profit maximizing and political maneuvering are also at work to defy simple, rational, apolitical laws of supply and demand. Moreover, what one views as "development" and what is best for the "whole" may differ from another, based on ideological, ethical, class and gender perspectives, all of which to a large extent shape the way we perceive and define development problems and their solutions.

There is a third piece in the literature, again from the late 1980s, by Nigel Twose, entitled "European NGOs: growth or partnership?" which also grapples with IOR development issues. Twose's essay addresses a number of dilemmas facing European NGO donors forming "partnerships" with African NGOs. The issues that Twose raises are fundamental to this research since they address NGO relations with donors and African governments. Twose argues that foreign donors and NGOs must recognize their role in development within the broader international political and economic context in which they operate.

With regard to African NGOs' relations to European donors, Twose argues that linkages go beyond the boundaries of those organizations. They extend to the governmental and economic policies of European home governments to which foreign donor funds are tied. Twose also makes the case that African NGO development projects conceived and developed outside of the African state development strategy have little chance of success, since those endeavors may be perceived as attempts to mobilize political opposition against the state. He questions how development relationships with African governments should be nurtured, and by whom—African NGOs or foreign donors? Twose also raises the gender issue, alleging the difficulty of "hand[ing] over all decision-making power to structures which do not involve women in any significant way and do not intend to." Tying all of this together, Twose questions the best way to resolve all of these issues: "by dangling a financial carrot around, or by approaching them in a spirit of genuine partnership?"[39]

Finally, Martin de Graaf, who focuses on local and national Zimbabwean NGOs, argues that NGOs can only be effective when they know, appreciate and can influence their wider environment. He posits that NGOs' success or failure depends on their ability to influence the environment optimally and to appreciate outside forces correctly. De Graaf proposed a very useful model of three concentric circles created by William E. Smith, Frances J. Lethem and Ben A. Thoolen to represent organizations and their environment.[40] This model is presented in Figure 2.1.

The innermost circle comprises those factors that are *managed* by the organization—the NGO—such as staffing, budgeting and setting objectives. The second circle represents those factors that can presumably be *influenced* by NGOs through persuasion, lobbying, patronage and exchange. These factors include government institutions and donor agencies. The third and widest circle is that part of the NGO environment that can only be *appreciated*. That is, it cannot be changed, influenced or controlled by the NGOs. It includes major national political structures and macro-economic systems, national and international. The boundaries between the circles are permeable.

De Graaf's presentation of the problem and the model he proposes for understanding the problem are quite insightful and thought-provoking. His application of the model is disappointing, however, in that, other than making a few token references to Judith Tendler's analysis of the risks and disadvantages of external financial dependence, he restricts his discussion of the economic environment influenced and appreciated in circles 1 and 2 to

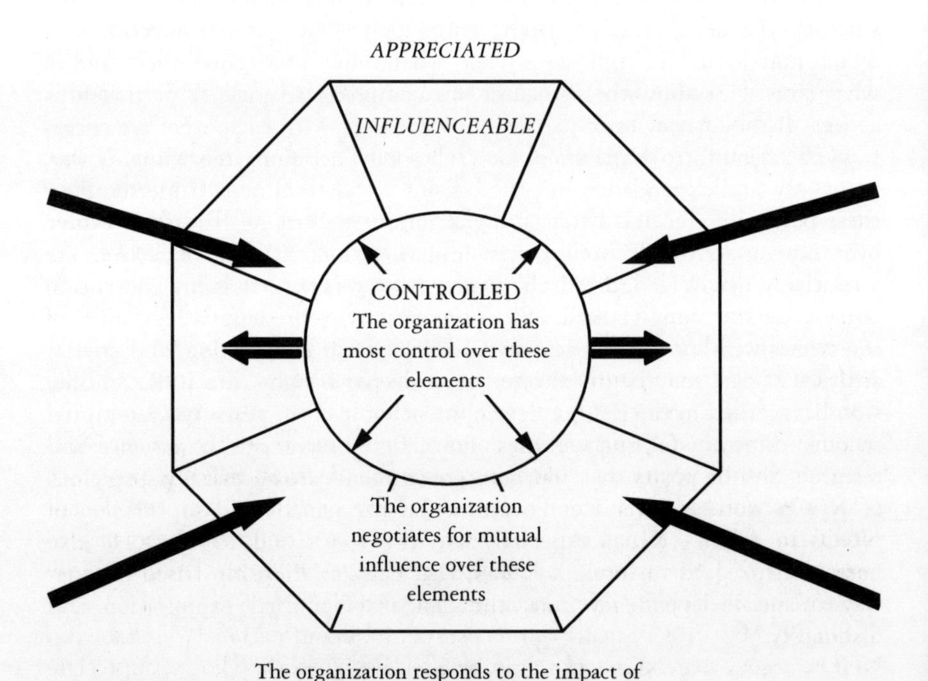

Figure 2.1 An organization's relations to its environments
Source Smith, Lethem and Thoolen (1980)

Zimbabwe only, and does not tie in the international dimensions of NGO funding.

Unlike Garilao, de Graaf pays a significant amount of attention to the relationship between NGOs and government. At one point he argues that NGOs linkages with government are weak, yet in another context he cites instances in which the government's behavior was influenced by NGOs. He makes the very important point, however, that NGOs in Zimbabwe see government not as a resource to be used but as a political authority to which they must be deferential. With regard to women, whom he mentions only briefly, de Graaf observes that women's NGOs make no attempt to (re)shape institutions which could affect the position of women.[41]

All of these works from the late 1980s by Garilao, Hirono, Twose and de Graaf raise important issues that are not yet satisfactorily resolved in the politics of development cooperation. The relevant studies of the 1990s, as previously discussed, thus far engage some of the issues that these authors raised, albeit with the noted exclusion of foreign donors. Overall, the studies from both decades offer valuable insights into IOR, but their major shortcomings are that they either do not 1) systematically grasp the entire network of all the organizational partners, including foreign donors,[42] or 2) examine the depth and politicization of the partners' interrelatedness in which they jockey for "who gets what, when, how" and "how much" and at what costs.[43] Studies which include the comprehensiveness of partnerships as well as the intricacies of partners' involvement with each other are necessary in attempts to understand how and why decisions regarding power, autonomy and dependence in partnerships are determined. Understanding these power differentials between organizations will more than likely evolve over time, as Southern NGO partnerships with transnational connections are a relatively newly recognized phenomenon for research. It is my contention that systematic longitudinal field observations, representative samples of interviews with *all* of the partners in development cooperation, and critical analyses of data may begin to offer some deeper insight into IOR. Smillie, who has worked in the field of development for over 30 years, has attempted to offer direction for such studies. From his observations, experience and research, Smillie posits that Southern governments are necessarily suspicious of NGOs, noting a trend of governmental harassment and suppression of NGOs in Africa—Kenya especially—in the 1980s and 1990s.[44] He also notes, with regard to foreign donors, that "any relationship based on one-way financial dependency runs the danger of mutual exploitation and dishonesty."[45]

DEFINING NGOS AND LINKING THE LITERATURE ON NGOS TO FOREIGN DONORS AND AFRICAN GOVERNMENTS: TWO SCHOOLS OF THOUGHT

Only in the last decade has the role of NGOs from the South in the development process come to the fore in social science literature. Consequently, very few empirical investigations exist, although the scholarly works are increasing in number.[46] This small but growing scholarly literature on NGOs reflects a relatively new recognition of the need for systematic understanding of the importance of the role of NGOs and the trilateral development partnership in the development process. Moreover, the current literature indicates that the nature of the relationship between NGOs and their partners in development has not itself been rigorously and sufficiently investigated. "What are NGOs?", "what exactly do NGOs do?" and "what is their relationship to government?" are questions still posed by scholars. Their answers are frequently ambiguous. Therefore, much remains unknown about the actual functioning and nature of NGOs in the development process vis-à-vis the nature of their partners in development.

DEFINITIONS—NGOS, FOREIGN DONORS AND GOVERNMENTS

NGOs

Acronyms, lexicons and varying imprecise definitions convolute an already complex area of investigation, so that it is necessary to spell out as clearly as possible the definition of NGOs in Africa.[47] Alan Fowler, in "NGOs in Africa, naming them by what they are," argues that NGOs cannot be defined as simply *not* being governmental.[48] That is not enough—there must be criteria which an organization must meet to qualify as an NGO and to distinguish itself from other organizations. Twose warns against Northern-imposed criteria to define Southern NGOs, for there is the risk of cultural imperialism in trying to make Southern NGOs mimic Northern NGOs.[49] There is also the risk of the Northern academic and practitioner community assuming that most African NGOs are local branches or affiliates of foreign-based organizations.[50] Although some may be, the vast majority of African NGOs spring from African roots. That is, some have indigenous roots and were formed in the early nineteenth century before the advent of colonialism on the continent. Other NGOs were formed specifically to challenge colonialism in the 1950s and 1960s—officially as well as underground; some were formed after the 1970s in the spirit of self-help and self-reliance, as in Kenya, for instance.[51] And, more recently, some have emerged continent-wide in response to worsening social and/or political and/or economic conditions. A strong case can be made that, comparatively, this last group has had the most profound development impact.

The Courier, a development journal published in Brussels, in a special issue on "Development NGOs" offers the following most useful definition and criteria that an organization must meet in order to be classified an NGO:

1 It should be autonomous. That means, it should not depend substantially on the state for its funds (although it may receive a portion of its funding from public contributions), nor should it be beholden to government in pursuance of its goals;
2 It should operate strictly for non-profit. That is, the funds it receives must be destined solely for its projects.
3 The majority of its operating funds must come from voluntary contributions;
4 In countries in the South, it should be created and operated by the nationals of the developing country of its origin, and it must work for that country's own development.[52] Theoretically, Southern NGOs may also operate in other countries towards the development of those countries as well.

Northern NGOs share this definition with Southern NGOs, except for criterion 4. For Southern NGOs, this definition is a malleable definition, often molded to meet the needs and requirements of Northern NGOs or their governments, thus at times undermining this criteria and mocking the notion of "NGOs." [53] *The Courier* recognizes the broadness of this definition and warns against other types of organizations in the South that may disguise themselves as NGOs. Some NGOs are involved in development activities, though involvement in development is not mandatory for an organization to be classified as an NGO. Those that become involved in development, specifically and exclusively, are sometimes but not consistently referred to as non-governmental development organizations (NGDOs).

Southern NGOs—specifically, in this case, African NGOs—may be local or national. The local/national dichotomy is suitable for the purpose of study, although it must be acknowledged that other more complex classifications exist but are not useful in this case.[54] At the local level, African NGOs are community-based and are run by the local members. They may or may not be linked to a national NGO. African national NGOs tend to have countrywide memberships and/or have countrywide operations (although this is not always the case), while they are based in capital cities. They have executive, administrative and sometimes professional staffs. They may or may not have NGO branches at the local level. National NGOs and local NGOs may or may not be assisted, both financially and technically, by bilateral and/or multilateral and/or private aid provided by foreign donors.[55]

Foreign donors

Foreign donors that become involved in development in the South may include the following: Northern governmental organizations (GOs), international governmental organizations (IGOs), international non-governmental organizations (INGOs), non-governmental organizations (NGOs) and businesses. Businesses are an anomaly and perhaps even a contradiction as partners in development cooperation for reasons which will be discussed later. Definitions of various foreign donors are as follows:

Northern governmental organizations (GOs) are organizations that are

1 created by legislative or executive measures of governments of organizations' origin, and/or
2 are funded by governments of organizations' origin.

GOs include organizations such as USAID, the US Peace Corps and the German Technical Agency.

International governmental organizations (IGOs) are organizations established by agreements outlining obligations between and among governments. IGOs include the World Bank and the International Monetary Fund.[56]

International non-governmental organizations (INGOs) are international organizations that are not established by intergovernmental agreements. They must meet the following criteria:

1 They must be international in character. That is, they must have intentions to cover operations in at least three countries;
2 There must be participation, including full voting rights, from at least three countries. Voting power must not be controlled by one national group;
3 A constitution must provide for a formal organizational structure and continuity of operations, giving members the right to periodically elect a governing body and officers. There must be a rotation of officers among various member countries;
4 There must be permanent headquarters for the organization that rotate at designated intervals among the various member countries;
5 Substantial contributions to the budget must be made from at least three countries. There must be no attempt to make profits for distribution to members;
6 An organization may have relations with other organizations, but there must be evidence that the former leads an independent life and elects its own officers.[57]

INGOs include OXFAM and the Salvation Army.

NGOs, both Northern and Southern types, were previously defined. Northern NGOs, which are not international, include organizations such as Africare. Southern NGOs include organizations such as the Green Belt Movement (GBM) in Kenya. MYWO also claims to be an NGO.

Mario Padron makes a very thought-provoking argument with regard to conceptualizing and problematizing NGOs. He argues that, fundamentally, the concept of NGO is inappropriately used to connote non-governmental organizations, namely the ones from the South.[58] He argues that development organizations from the South should be identified as NGDOs, and that they should be distinguished from grassroots sector organizations (GROs)—NGDOs are local or national, and GROs are grassroots. Padron's points are well made, as "NGO" is generally used to refer to an amorphous set of organizations ranging from religious institutions to trade associations to cooperative societies to international agencies. Often, little distinction is made between international, national and local NGOs, particularly "on the ground" in the development cooperative relational environment of development implementation. This "generic" use of NGO informs the scholarship and further contributes to the conceptual muddle and confusion surrounding NGOs. Padron's call for conceptual clarification is timely. Yet another problem to be raised is defining an "NGO" which has coterminous boundaries with its home government, like MYWO at certain points in its history. Is it an NGO? This discussion will be engaged more exhaustively in later chapters.

Businesses are private profit-making enterprises. They are not NGOs or GOs. They may be local, national or international. The business that became involved in development partnerships with MYWO during the period of my research, as well as during an earlier period, was the Kenya branch of Coca-Cola, the transnational corporation (TNC), otherwise called a multinational corporation (MNC).

That a national branch of a TNC would be a foreign donor in a development cooperative partnership is perplexing, as stated earlier. Many development scholars, armed with ample evidence, would argue that businesses, by their intrinsic nature, are contrary to cooperating in development "partnerships." The reason is that TNCs have historically been a cause of underdevelopment in the South. In addition, compared to IGOs, INGOs and NGOs, businesses are least prone to espouse humanitarian rhetoric and/or a commitment to development assistance.

In this case study, however, the business Coca-Cola–Kenya was considered by MYWO and also by itself to be a "donor" to MYWO in a kiosk income-generating project for women. Coca-Cola–Kenya provided the material assistance to MYWO in a cooperative development project which involved MYWO, Coca-Cola–Kenya and the Nairobi City Commission. Coca-Cola's rhetoric indicated that the goal of this project was humanitarian as well as a marketing strategy and profit-seeking for Coca-Cola. Women who were issued kiosks were to sell a stipulated quota of Coca-Cola bottled drinks monthly, among a range of products for sale. The Coca-Cola products were to be purchased by the women at wholesale prices from Kenyan distributors. The women were prohibited from selling Coca-Cola's competitors' products.

Together GOs, IGOs, INGOs, NGOs and businesses comprise the "foreign donors" for this study.

African governments

Governments are defined as the occupants of public office who make decisions at a particular time. They may be viewed as the policy actors of the state. The state is the organized set of institutions, associations and agencies claiming governing power over a defined territory and its population. With society, the state and the government of the day may engage in acts ranging from negotiation to domination.[59] The state and government institution that becomes involved most directly and actively in development cooperative partnerships is the bureaucracy. The bureaucracy represents civil servants in the various government ministries or bureaux. That is, bureaucrats from the lower rung to the top-level managerial staff become involved in development cooperation by providing technical and/or financial assistance to NGOs. They do this at the will of heads of governments, who may create environments ranging from cooperation to conflict with NGOs. African heads of governments are able to do this, as they often have *de facto* unchecked and unlimited state power to facilitate or quell development efforts. How the government engages in development in Africa depends largely on the regime type and the nature of the state. With regard to the government's approach to women's development activities, much depends on the head of government's ideological and personal views towards women and gender. The Moi government from 1978 to 1992 is the most important period in Kenya for this study, because it was during this time that MYWO's status as an NGO was officially changed, first to a GO in 1987 and then back to an NGO in 1992. It was also at this time that many in its relational environment began to question MYWO's autonomy and integrity as a women's organization. The ministries most actively involved in development partnerships were the Ministries of Culture and Social Services, Agriculture, Energy, Livestock, Health and Home Affairs and National Heritage.

AFRICAN INDIGENOUS NGOS AND THEIR PARTNERS: TWO SCHOOLS OF THOUGHT

The nexus between NGOs, foreign donors and African governments is indeed a complex, potentially fluid and volatile one. A review of the two schools of thought regarding their development cooperative relationship illustrates this, for there is dissension among scholars as to the role of indigenous NGOs, the contribution they actually make to development, and the nature of their relationship to their partners in development. Within the last 15 years, for example, there have emerged diverging perspectives, both descriptive and prescriptive, which I shall refer to as two schools of thought. These schools hold diametrically opposite viewpoints.

For example, the pessimist school, on the one hand, emphasizes the politicization of the relations between partners in development. The optimist school, on the other hand, is more naive about international power relations between states, organizations and their subsequent effects. The optimist school does, however, also take into consideration—though not as critically as the pessimist school—the *politics* involved in development cooperative partnerships.

INDIGENOUS NGOS AND DEVELOPMENT PARTNERSHIPS: THE SCHOOL OF OPTIMISM

The first school of thought suggests that NGOs from the South become agents of development representing the interests of indigenous people. Peggy Antrobus argues that NGOs "can be seen as an expression and outgrowth of the ending of colonial rule, a symbol of people's new-found confidence in defining their own needs and priorities and in taking responsibility for addressing these."[60] Pradervand and Bernard Lecomte agree that NGOs become the leaders of partnerships in the development process—taking responsibility for meeting the needs of their constituency and setting their own agenda.[61] Generally, it is believed that NGOs have demonstrated the capacity to design and implement development programs, using innovative approaches without the governmental hassles and bureaucratic red tape, which actually reach the people at the grassroots.[62]

This school generally supports its optimist view by citing instances in which NGOs have demonstrated potential for sensitive, successful and self-reliant development in Burkina Faso, Mali, Kenya, Senegal and Zimbabwe.[63] Generally, NGOs are believed to assume charge and direct their partners in development—foreign donors and the government—implementing NGOs' own development agendas with the financial and technical assistance of their partners, but not becoming beholden to them. More and more, from this school's perspective, NGOs are viewed as playing a more effective role in responding to Africa's crises and in building strategies which will bring substantial development to the continent.[64]

Garilao supports this view of NGOs by asserting that NGOs are emerging as advocates for self-reliance who will, by themselves, bring about significant policy and institutional change. He calls for NGOs' partners in development to aid in defining an agenda without imposing their own agenda, to be "facilitative not directive."[65] Fernand Vincent, founder of a Geneva-based foreign NGO, makes the same argument in a different manner: "Western NGOs should efface themselves more and more and let this movement come to fore. The future belongs to [Southern NGOs]—if they manage to acquire the needed means and know-how and a real understanding of economic forces."[66] Generally, this school takes the stand that NGOs leading development partnerships perhaps hold the key to Africa's future. David Steel, Program Coordinator of the World Bank Administration, has described NGOs in this way: "NGOs are already substantial contributors to the global and local dialogues on development issues, are substantial providers of human, financial and resources in kind, and are well positioned to respond to the challenge of the twenty-first century."[67]

INDIGENOUS NGOS AND DEVELOPMENT PARTNERSHIPS: THE SCHOOL OF PESSIMISM

The second school of thought is less optimistic, however. This school was born out of the belief that NGOs are "oversold" since their presumed effectiveness in development partnerships is not as positive as the school of optimists tend to think.[68] In this school, it is argued that NGOs may become mere vessels through which their partners gain support for their own programs with no real regard for NGOs themselves. For instance, Hendrik van der Heijden has discussed threats to the autonomy of Southern NGOs due to their excessive dependence on foreign financial assistance resulting in the subordination of NGOs' own "internal" agendas. He uses figures from Jacob Mwangi, who demonstrated that in 1986 the dependence rates of sub-Saharan African NGOs exceeded 90 percent, which van der Heijden positively correlates with weak internal capacity for development program formulation.[69] De Graaf further argues that "there are no immediate prospects that this tendency will abate" for NGOs are not "systems on their own" but are integrated into a wider and more complex political and administrative environment in which they have "limited influence and even less control."[70] Helen Allison uses an example to illustrate this point and argues that: "British aid is moving away from helping the poorest people and is increasingly being used to sell British exports and to further the government's foreign and economic policies."[71]

Kabiru Kinyanjui further speaks to the involvement of foreign donors in the African development process. He argues that foreign aid is increasingly being used to 1) control international trade in the favor of industrialized nations, 2) depress agricultural commodity prices, and 3) prescribe the "bitter-medicine" conditionalities of the World Bank and IMF. In essence, Kinyanjui argues that foreign aid is being used as a foreign policy tool.[72] Sam Kobia evaluates the functions of foreign donors in Kenya and lends support to Kinyanjui's views. He believes that NGOs and foreign donors "run the risk of being pawns in the chess game of manipulating the poor in order to perpetrate unjust global socioeconomic relationships." He argues that, through operations disguised as development "partnerships . . . the rich North reaps handsome benefits while the South is always the loser."[73] According to Kobia, development partnerships are merely token mechanisms promoted by the rich North to forestall any serious attempt to redress the unjust international economic order. Charles Elliot, former director of the INGO Christian Aid, speaks to this possibility. He states, "Aid can be used as a way of diverting attention from much deeper economic injustices which perpetuate poverty on a far greater scale than the most generous aid programme can relieve."[74]

Yash Tandon, perhaps the most critical of the pessimist school, argues that Northern NGOs may in fact be new conduits for (re)colonialization of Africa from the grassroots in the 1990s. He bases his argument on several factors. First, he charges that foreign donors in Africa are a "secret lot." Not much is known about them and their origins. Second, although much of foreign donor monies comes from government, institutions through which this government money is funneled are called "non-governmental." Advocates of the pessimist school, like Tandon, tend to collapse foreign aid into one category in their critiques. Kobia, for instance, argues, as many others suggest, that governmental aid invariably permeates non-governmental aid, hence making distinctions between governmental aid (multilateral and bilateral) and private aid moot in the politics of development cooperation. Third, Tandon states that foreign donors' development agendas are decided in the North and not with their target populations in Africa. Fourth, many Northern NGO donors in America were founded at the height of national anti-Communist paranoia and have since been involved in counter-revolutionary activities in Southern Africa, particularly Mozambique, Zimbabwe and South Africa. Fifth, Northern NGOs make an issue of violations of political rights by African governments, but they do not make an issue of violations of economic rights by the international capitalist system which exploits peasants' labor. Last, Tandon posits that many of the Northern NGOs come from countries which plundered Africa, through slavery, colonialism, then neocolonialism. The North's next step, Tandon argues, is colonialism on a different level—the grassroots.[75] Southern NGOs may be their avenue.

From the literature grounded in the pessimist school it appears that, in this trilateral relationship, African governments are in very precarious positions. They want the financial resources that NGOs are able to attract and secure from abroad as well as the political patronage they are able to acquire through allocating these resources locally, yet they do not want the political challenge from the independent power base that NGOs often build.[76] Michael Bratton demonstrates that NGOs are among the few organizations in developing countries to maintain a direct presence among the grassroots, and on that basis, he argues, they may be perceived as threats to the state. David C. Korten and L. David Brown, like Bratton, further take the position that popular mobilization, participation and pluralism are part and parcel of NGO's development successes, and are essential elements in what makes autocratic governments uneasy.[77] They state that the very existence of NGOs promotes a democratic political culture which 1) encourages and supports popular participation, and 2) presents serious contenders for competing ideas and alternative political leadership with legitimate grassroots support. Milton Esman and Norman Uphoff argue that such a reality, as perceived by the government, may be detrimental for NGOs and their potential for development success. Hence, they argue that most successful NGOs "enjoy the support or at least acquiescence of government and are linked to services and resources that originate in the state."[78] For this reason, they argue that absolute autonomy of NGOs is not desirable for rural development. If a government feels its hegemony is threatened, as Bratton argues, it may resort to one or more mechanisms of control, including: 1) monitoring and registration, 2) coordination, 3) cooptation, and 4) reorganization, dissolution and imprisonment.[79]

Ndegwa best demonstrates the dialectic relationship that African states have with African NGOs, using Kenya as a country case study, and the Green Belt Movement and Undugu Society as comparative NGO case studies. Ndegwa's basic argument is that African governments view NGOs as both socio-economic assets and political contenders, as partners in development and competitors for state power simultaneously. This, he states, assuredly places NGOs and states on a collision course.[80] Ndegwa further challenges the earlier assertions of Korten and Brown and Bratton that NGOs are necessarily pluralizing, democratizing or strengthening of civil society. He convincingly makes the case that there is nothing inherent in NGOs or larger civil society that necessarily makes them "opponents of authoritarianism and proponents of democracy."[81] He argues that for NGOs to be a democratizing force they must be operatives in a larger political context in which there are two critical externalities: 1) a social movement, and 2) political opportunity.[82] Ndegwa further juxtaposes the African NGOs against foreign donors and argues that it is NGO dependence on foreign donors and not African governments' control of NGOs that serves to diminish NGOs' political opportunities, as well as their potential to pluralize civil society and engender democratization.[83] Because of this, Ndegwa sympathizes with African governments' dialectic views towards NGOs and posits that governments have a "rather" legitimate suspicion of NGOs.[84] At the same time, he remains very critical of the Kenyan government's passage of the NGO Coordination Act of 1990, which was intended to control the activities of NGOs.[85]

These issues, as presented, describe the main points of departure between the two schools of thought—pessimism and optimism—regarding the role of indigenous NGOs in development vis-à-vis their partners—foreign donors and the African governments. It is important to note that there are perspectives that do not fit in either one of these schools of thought.[86] However, these two schools of thought on the nature of the interorganizational partnership represent the current competing development cooperation paradigms with regard to partnerships. The dissension and competition between these schools of thought underscore the importance of understanding the linkages and effectiveness of development cooperation between organizations, as North and South come together in NGO, foreign donor and African government partnerships.

WOMEN, NGOS AND IOR: A MISSING LINK IN DEVELOPMENT THEORY AND PRACTICE

Glaringly absent from the past discussion of NGOs and IOR is the inclusion of women. One would imagine that in the literature on indigenous NGOs particularly, there would be a wealth of information on women, as women comprise the bulk of NGO memberships in Africa. This, however, is not the case. Assessing the proliferation of NGOs on the African continent in the late 1980s, Pradervand noted that NGO memberships are predominantly comprised of women, even when they do not appear at the heads of organizations. For example, in the Sahel, Pradervand estimates that women make up between two-thirds and three-quarters of organized peasants—in Kenya up to 90 percent.[87] He further explains that women make up the overwhelming majority of NGO members because women have been estimated as the majority of rural inhabitants and farmers who produce over 75 percent of Africa's food, while in some areas it is men who are more likely go to the city to look for work.[88]

Despite these facts, many studies point to the indisputable exclusion of women from development planning and development scholarship, beginning in the 1970s with Ester Boserup's study which tied women and development concerns together in an international context.[89] In spite of noting this exclusion of women, and in spite of noteworthy attempts to pay closer attention to women's roles in development,[90] nearly 30 years after Boserup's study the scholarship and literature on women remain comparatively marginalized on several fronts. The first of these is that much of the literature remains effectively separate and apart from other works on development. For instance, much of the research on women in development tends to be in-house evaluations by international organizations such as USAID (special section on women in development), or special series papers on "women and development."[91] This in itself is not problematic. What is problematic, however, is that these studies are kept for the most part in-house and consequently remain separate and apart from "mainstream" development literature. Second, the literature on women tends to remain within the same circle of scholars. That is, it is produced and consumed by the same group of scholars, mostly women, rarely making its way beyond the margins of this group to the larger development audiences or to the women it often studies as research subjects or recipients of aid.[92] Although this scholarship has contributed to more attention and assistance from the North being channeled to women's projects, one of the major criticisms is that rarely is this assistance used to initiate *appropriate* development action together *with* women from the South. Sue Ellen Charlton argues that we may one day conclude that the greatest impact of the focus on women in development is not the betterment of women's life chances in tangible ways, but instead the financing of a collection of a wide array of studies relevant to development in general.[93]

MALE BIAS IN DEVELOPMENT THEORY

The marginalization of women and development scholarship on women is a direct consequence of male bias in development theory. Although less of a bias exists today as compared to a quarter of a century ago, male bias continues to exist at the theoretical level, and it spills over into the formulation and implementation of development policies and practices.[94] Just as IOR has been marginalized by mainstream political science, women have been marginalized throughout the evolution of development theory since the 1940s. Development theory over time has generally been created by men from the North, and when these theories have attempted to paint a human face they have been biased, whether intentional or not, towards the North first and towards men in "developing" societies in the South second.

From modernization theory of the 1950s and 1960s, which stressed tension and conflict between tradition and modernity and assumed that societies would choose modernity and thus "pass" from the former to latter, there has been male bias. For instance, modernization theory measures the extent of "modernity" or "backwardness" of a society based on variables that are, by their nature, biased towards men. It measures: 1) performance and stability of political systems, "systems" being patriarchal and male-dominated in their conceptualizations; 2) quality of the politics, "politics" defined as taking place in the "public sphere" which is the traditional domain of men; 3) percentage of literate persons in the population, a reflection of gender bias as men were (and still are) more likely than women to be formally schooled; 4) percentage of workforce in non-agricultural occupations, a reflection of men since women's primary occupation was (and still is) that of farmer; and 5) percentage of urban population, again a reflection of men since most women remained on the farms in the rural areas to work while more men traveled to the urban areas.[95]

In addition, underdevelopment and dependency theories of the late 1960s and early 1970s have further focused on men. These theories focus on extraction of resources from the periphery by external forces at the center. This extraction serves to perpetually bankrupt the periphery, which includes the African continent. Moreover, center-periphery linkages serve to create a comprador elite or a local bourgeoisie, which emerges usually as a male elite group that benefits from this extraction and protects the interests of the center in the periphery. This group also protects male–state interests in the periphery.[96] Women from the periphery, as a group, are the most marginalized in this equation. The works of Shelby Lewis, Beverly Lindsay, Charlton and Barbara Rogers provide ample support for my point, and this case study will further reiterate it. Even the more insightful theorists of Africa's conditions of underdevelopment and dependence, such as Walter Rodney, Samir Amin and Colin Leys, did not explicitly address the plight or the potential of women in the periphery. Immanuel Wallerstein's world systems/dependency work did not address the issue of women either, until an edited work in 1984.[97]

The statist approaches to development theory in the 1980s also focused on men. In that African states began to look more introspectively and accept the brunt of responsibility for development (or lack thereof) in this period, such states focused more directly at this time on urbanization and industrialization as the catalysts for development. Development in this vein was projected to come from male urban working-class labor, although the state lacked the skilled (Western-styled) male labor power in the urban centers as well as the appropriate technology to industrialize. These statist approaches ignored women, who live predominately in the rural areas, who are agriculturists primarily, and who are the bulk of the continental labor force.[98] Political development scholars, including Samuel Huntington, Gunnar Myrdal and Leonard Binder from the North, have indirectly perpetuated this bias against women, as they have separately supported statist theories since the 1960s. Their writings suggested that 1) the state should harden its stance against popular mobilization and political activity as state institutions were still "weak;" and 2) the state should, moreover, draw "peripheral" populations into national development, *only under the terms of the state.*[99] More important than their contribution to development theory, at this point, might have been their justification for statist one-party regimes in Africa.

The political economy approaches to development of the late 1980s and early 1990s have also been biased against women. They have focused on replications of the industrial production and export schemes of the Asian economic miracles—particularly those of Asia's four tigers, Taiwan, Hong Kong, Singapore and South Korea—as they correlate development with economic growth.[100] Despite the exploitation of women's labor being part and parcel of these economic growth models, they are endorsed by development theorists of this school as "exportable" development models. Much of the bias against women stems from the oppressive myths that women are necessarily docile, dexterous and able to swiftly complete painstaking tasks. Women are also perceived to be the most reliable and least demanding of all workers, as job options for them in the formal employment sector are few to nil. Women laborers in export industries are the least regarded and most exploited of all laborers under these development models.[101]

Finally, the policy process approach, as R. William Liddle argues, interjects the mandate to formulate and implement the "right policies" in development theory.[102] This approach is among the newest for unraveling the perplexities of non-development in this decade. Liddle urges scholars to consider the political calculations that politicians necessarily consider in the formulation of development policy. To extend Liddle's argument, I contend that development theory must consider the calculations of not only male but also female politicians, in both governmental and non-governmental spheres. Moreover, development theory must consider the actions of male and female politicians in the development policy process juxtaposed against the pressures politicians face from internal and external sources, as well as against the sincerity (or insincerity) of individual politicians' desires for national development. I further contend that policy formulation and implementation in Africa must be considered in the context of 1) the coexistence of the primordial public and the civic public,[103], and 2) the separate political cultures and "cultures of politics"[104] of men and women, as well as 3) the pressures from the international capitalist economy. This holistic interactive policy approach which Liddle suggests, when coupled with a gender perspective which I am suggesting, may offer women the recognition and voice in development theory which hitherto they have not had—that is, if scholars and practitioners are truly committed to formulating policies devoid of male gender bias.

With the exception of this modified policy process approach, almost all of the other theories and approaches to development recounted here share urban, elite and industrial biases which are tantamount to male bias in development theory as it relates to Africa. Generally, development theory ignored the rural populations, of whom many are women, until the Women in Development (WID) decade. In spite of women's recognition, as a consequence of the WID decade, there is still bias in favor of men.[105]

CURRENT CHALLENGES TO MASCULINIST DEVELOPMENT THEORY

Catherine Scott and Jane Parpart separately bring the exclusion of and bias against women in development theories and approaches to the fore in two recent works. Scott reinterprets the canons of development theory, specifically modernization and dependency theories, and demonstrates that both are fundamentally hegemonic in their premises and perpetuate "gendered" meanings of development and dependence.[106] She argues that modernization theory correlates masculinity with modernity and femininity with tradition, while suggesting that development requires the emergence of a modern, rational *man* who is receptive to new ideas. This, Scott continues, requires being in opposition to a feminized, traditional household which pits women as closer to nature and as unreceptive to new ideas, and justifies a sexual division of labor in gendered spheres of life—mental labor assigned to men in the public sphere, and physical labor, productive and reproductive, assigned to women in the private sphere.[107] She further argues that despite dependency theory's more accurate description of women's lives as workers—producers and reproducers—dependency theory "also tends to view women's labor as natural and bound to the household, while the public realm is a privileged location for revolutionary activity"[108] for men. She makes it clear that, despite the revolutionary rhetoric of dependency theory, the binary opposition of tradition and modern, backward and progressive, and, concomitantly, feminine and masculine is maintained.[109] These are false dichotomies.

Parpart illuminates the degree to which these dialectics are maintained, as well as the manner in which they continue to exclude women in development approaches, particularly in WID, women and development (WAD) and, to a large extent, even gender and development (GAD) discourses.[110] She argues that, despite their rhetoric, development theorists and practitioners—mainstream and alternative—have continued to embrace the modernist paradigm of development. This paradigm purports (sometimes by women from the North) that development for women in the South virtually means throwing away the "traditional" (the indigenous) and accepting the more "modern" (the Western).[111] It further maintains a bias against women from the South by depicting them as overburdened helpless victims, and hence objects and recipients of development; while it is biased towards men from the South who receive most of the consultancies in the South, and who often ignore the salience of gender because of bias against women.[112] Parpart correlates the modernist paradigm, as it relates to development approaches in the South, to colonial discourse. She argues that the emphasis on women from the South as vulnerable helpless victims is a mere throwback to the racist and ethnocentric colonial rhetoric of "uplift[ing] the natives."[113]

The works of Scott and Parpart challenge the legitimacy as well as the ethics of current development theory and approaches. Their works also weaken the assertion of David Booth, who argues that by including gender analysis in development research, current paradigms have become less limited than they were prior to the early 1980s.[114] To some extent, this may be the case, but Scott and Parpart have demonstrated that the mere inclusion of gender analysis within the modernist framework does not challenge the historical masculine orientations of development theory and practice. Perhaps it is at this impasse that we may begin to address the source of the problem.

Caroline Moser recommends gender planning for development to get beyond this impasse. Gender planning for development she defines as the means to emancipate women from gender subordination and to achieve gender equality, equity and empowerment.[115] Moser recognizes the importance of women's interactions with and in NGOs, as well as women's organizations' vulnerabilities to possible exploitation by foreign donors and states in trilateral development partnerships.[116] Although Moser's attempt is to interject gender more centrally in development in practice, it does not promise to shift the current development paradigm, for it remains fundamentally modernist. It merely attempts to integrate women into development as currently practiced. However, as women, the state and foreign donors come together in this era—interlinked as a new institutional interorganizational phenomenon on the cutting edge of development implementation—development theory will be forced to grapple with women, gender and NGOs "front and center" in its discussions, debates and theory reformulations.

THE NEXUS BETWEEN WOMEN IN AFRICA AND WOMEN GLOBALLY

It is an understatement to say that women's life options are directly related to their relationship with and access to powerful men.[117] Robert Fatton highlights this phenomenon in the African context. He argues that "women's access to political and economic resources has been severely constrained by pervasive and overwhelming patterns of male domination."[118] In relatively all aspects of life, he demonstrates that women's opportunities and status are determined by their connections to men—fathers, husbands and brothers. Using Ghana as a case study, Akosua Adomako Ampoto further notes that gender identity in Africa can often be the basis for objectification, blame and persecution, when one is female.[119]

From a global perspective, in most if not all societies women's opportunities are determined by men to varying degrees.[120] Women as a group worldwide are politically, socially and economically dependent on men. The reason is that, with respect to class, race, ethnic and country of origin differences and their corresponding privilege differences, "women in general have little or no formal, institutionalized power at the local, national, and international levels in comparison to men."[121] Moreover, the increasing interrelatedness of these three levels—local, national and international—further reinforces the political power differences between women and men. Thus, women from the South are likely to have the least power of all, as men from the South have less political and economic power than men from the North.

John Stoltenberg and bell hooks separately argue that men, as a group worldwide, dominate women by establishing patriarchal control over women through a process which Stoltenberg describes as male bonding. He provides the following definition: "Male bonding is institutionalized learned behavior whereby men recognize and reinforce one another's bona fide membership in the male gender group and whereby men remind one another that they were not born women. Male bonding is political and pervasive."[122]

hooks agrees, but points out that, even though men have political privileges and power over women simply because they belong to the gender group of men, it is possible for men to change. For example, she argues that when men raise their political consciousness and realize the oppressive system that interlinks classism, racism, capitalism and sexism, they come to the realization that although women are more oppressed, men are not free either.[123] hooks goes further than Stoltenberg and states that it is not individual men who are the real enemy of women or the cause of the power differences between men and women. Instead, hooks argues, it is a ruling-class patriarchal system which perpetuates itself through the socialization of men to engage in and encourage men's sexist and oppressive practices against women. Its ultimate aim is the control of nations, states and economies.[124] With a view to women globally, Shelby Lewis demonstrates that this patriarchal system of which hooks writes is in fact a global patriarchy, bonding men and oppressing women worldwide.[125] Despite the magnitude of the patriarchy, hooks' argument strongly suggests and recommends that male bonding be unlearned.

Maria Nzomo demonstrates that Kenyan men are part and parcel of this global patriarchy which oppresses women. She highlights Kenyan ruling-class men's successful attempts since independence to systematically exclude women from the formal political and economic structures of the state, sanctioning only the censored and supervised presence of hand-picked government-and-male-identified women within their ranks (albeit on a lower level).[126] Generally, politics and economics in Kenya, as in most other countries, are considered men's affairs.[127] Concomitantly, the home and childcare are considered women's affairs. Nzomo proposes a challenge to these separate gender spheres, stipulating that politics and economics not be deemed the domain of "men only." Nzomo calls for women to "own" their organizations, to make their agendas political and to reorient their activities towards women's economic control.[128]

WOMEN, ORGANIZATIONS AND IOR

What we know about organizations, from Mike Savage and Anne Witz in a path-breaking study which attempts to mesh organizational theory and feminist theory, is that organizations are in fact "gendered."[129] Despite this fact, little work has been done on women in organizations, and even less on women's organizations in IOR, especially on women as political actors in NGOs. Nonetheless, recent scholarship on women in various types of organizations is instructive and has some applicability for understanding women's organizations in IOR studies. For example, Celia Davies, in a study of women and nursing organizations, found that gender is a "relations quality" in organizations and gender relations are power relations.[130] In addition, Karen Ramsay and Martin Parker argue that, within organizations, "women are excluded as equal organizational participants by patriarchal structures and processes" while men as a group retain power and authority through legitimizing hierarchical organizational structures.[131] Rosemary Pringle challenged this structural determinist perspective and argued that men do not necessarily unilaterally impose male power on women. Instead, Pringle argued that gender relations are more a process involving strategies and counter-strategies of power.[132]

Ramsay and Parker provide the most relevant information from which we may conceptually understand a gendered organization in a patriarchal world. They state that "organizations are [mere] constructs with the interpretive resources that any culture provides . . . [thus] gender oppression is common to most if not all organizations."[133] Hence, based on this, it is unrealistic to expect that women in organizations and women's organizations in IOR would *not* be subordinated by the power differential which exists between men and women, and which is supported in their relational environment globally. Elise Boulding argues that this power differential has been an impetus for the creation of NGOs as alternative organizations for women. Writing from a historical perspective on NGOs globally, Boulding states:

> The phenomenon of the women's NGOs stemmed in part from the inability of women to get men to give priority to decentralism and non-violence, and in part from the fact that men could not perceive women as individual human beings in their own right, let alone as partners in major public enterprises.[134]

SPECIFIC REFERENCE TO AFRICA

Outside impact studies on women's contributions to community development, serious scholarship on women's organizations in mainstream development IOR literature in Africa is lacking. Moreover, in-house studies tend not to be available for public consumption. This is problematic considering the fact that IOR involving women's organizations has grown phenomenally on the continent in the recent past.

To date, women's organizational affiliation in Africa is exponentially higher than men's.[135] Subsequently, women's IOR affiliations are markedly higher than men's as well. Despite this organizational affiliation and IOR involvement, women's organizations' gains remain minimal. After a critical look at women and the state in Africa, Parpart and Kathleen Staudt argue, and Naomi Chazan concurs, that women remain outsiders in their relations to the state, severely restricted from voicing their concerns in the public political arena.[136] Generally, organizational affiliation offers them no real independent voice. Likewise, Nzomo argues that women's organizations become more exploited, oppressed and dehumanized when they interact with foreign donors, as donors treat them as a perpetually available pool of cheap labor.[137] Rogers underscores the validity of this argument as she points out that international organizations are not exempt from Western male bias in development.[138] Sally Yudelman, whose work concentrates on Latin America, but who makes more general observations as well, supports Rogers' point by demonstrating that some Northern donors have not done very well at integrating women into their own Northern organizations.[139] Harvey Glickman and Staudt posit that international organizations may in fact be part of the problem, allocating resources to men via men, while simultaneously being slow to respond to women's work and needs.[140]

At a very fundamental level, women's organizations in Africa operate in a relational environment in which they are particularly vulnerable to male domination and patriarchal ideologies, policies and practices that are fostered by either the state or foreign donors, or both. To state this differently, women's organizations are dependent, to varying degrees, on the patriarchal structures of the state and foreign donors for their continued existence and prosperity. This is not surprising since patriarchal traditions are deeply rooted both in the fabric of many African traditions[141] and in the politics of interorganizational relations, which serves to reinforce oppression against women as individuals and in organizations.

Nzomo argues that women are partly to blame, for women's passivity has allowed men—and hence states and governments—to succeed with the implementation and continuation of patriarchal policies and practices.[142] An argument by Staudt, however, demonstrates that the situation may be far more complex than Nzomo represents. Staudt argues that it is sometimes difficult for women to act on gender-based concerns consistently, since women also have concerns of family and other matters which may cross-cut gender concerns.[143] Janet Bujra accepts Staudt's contention, as she argues that to believe that women's organizations in Africa emerged to challenge the patriarchy is to begin with a false premise. Bujra states:

> The existence of women's organizations in Africa is not . . . unthinkingly to be equated with the existence of any specifically feminist consciousness, or any desire to transform the class or economic structures of postcolonial society. Women's liberation is disruptive in its challenge to male prerogatives; organizations such as these reinforce the status quo. They serve petty bourgeois class interests more than they serve women. To dismiss them out of hand because of this would be shortsighted, however. For, despite their primary significance as institutions of class control, such organizations, in bringing women

into communication with each other, can provide arenas of struggle within which women who are poor and subordinated can speak out and exert pressure on those who enjoy the rewards of postcolonial society.[144]

There are scholars who would challenge Bujra's argument, yet her point is well made. There are some cases in which women have resisted openly patriarchal control,[145] yet in most instances women have exited, albeit quietly, from the fringes of the political domain rather than challenge men and the state. This seemingly passive apolitical stance that women take may seem perplexing on the surface, yet when one digs deeper it becomes clear that "exit" is a very rational stance, and that its fundamental underpinnings are extremely political. That is, women's organizations and individual women who exit from formal involvement in politics often do so because they have contrived formulas for survival which maintain for them at least a limited voice and chances for success (however small) in a very limiting, patriarchal, oppressive and often misogynistic environment. By exiting, they aim to at least survive and at most prosper.

That women's organizations in IOR are confronted with and are dependent on male power in a patriarchal relational environment, nationally as well as globally, is an avenue of investigation IOR must undertake.

FRAMING THE TRANSNATIONAL RELATIONSHIP OF MYWO, FOREIGN DONORS AND THE KENYAN GOVERNMENT: BUILDING ON THE FOUNDATION OF IOR'S RESOURCE DEPENDENCY MODEL (RDM)

There is no explicit model in the literature which brings together IOR, indigenous NGOs and women. This study is one of the first to explore the interconnectedness of these areas of research using this framework. There is, however, an IOR model, the resource dependency model (RDM), which is capable of addressing, at least in part, the politicization of the transnational network linkages which exists when foreign donors, African indigenous NGOs and African governments come together as "partners" in development cooperation. The foundation of the RDM offers a starting point to begin to identify, assess and compare variable impacts of power, autonomy and dependence of the partners in development cooperation. This model must, however, be modified in order to address the scope and peculiarities of this study.

Specific to this research problem, IOR's resource dependency model appears quite applicable in that questions regarding the autonomy of MYWO,[146] as a Kenyan women's organization which claims to be an NGO, and its dependence on foreign donors and the Kenyan government in its relational environment are the foci of my project. Autonomy, relative power and dependence are the key concepts around which RDM revolves. Because

MYWO is a women's organization, RDM is an even more appropriate basis for analysis since women as a group, in a world of patriarchal power and politics, have less autonomy and power and are dependent on men for resources in their relational environment. Thus, for this study a gender component of RDM was created.

DEFINING THE RESOURCE DEPENDENCY MODEL

RDM has strong theoretical ties to the concepts of political economy, dependence and exchange in IOR. Moreover, RDM appears to have grown out of expansions of and challenges to Sol Levine and Paul White's exchange theory (1960–1) which served as the dominant model for analyzing and explaining interorganizational behavior for over a decade.[147] Levine and White define organizational exchange as "any voluntary activity between two organizations which has consequences, actual or anticipated, for the realization of their respective goals or objectives."[148] In response to Levine and White's propositions, other IOR theorists, in contributing to a definition of RDM, argue that organizations which lack the necessary resources—which they define as 1) recipients to serve, 2) resources, and 3) personnel to direct resources to recipients in order to achieve their goals and objectives—engage in exchanges with other organizations.[149]

One of the major challenges to Levine and White which was brought to bear on RDM was based on their conceptualization of "exchange." It has been argued that Levine and White's definition of exchange was so broad that it could include "*any* form of voluntary activity between organizations, [thus merely] rendering the term "exchange" synonymous with interaction."[150] Furthermore, it has been shown that Levine and White's definition of "voluntary" exchanges captures the "normal" aspects of exchanges brought about in the absence of power differences and dependencies.[151] From an RDM perspective, such a definition of exchange is apolitical, naive and unrealistic. It excludes interorganizational relationships that involve coercion and domination, where exchanges might be involuntary.[152] This is important to note, for when women's organizations are involved in male-dominated IOR, because of gender differences and subsequent power differences, there is a strong likelihood that exchanges embedded in the politics of coercion and domination may be frequent occurrences; hence the need for gender analysis.

Resource dependency theorists argue that "potential power is equivalent to the possession of scarce resources" and "organizations that acquire a monopoly over resources are able to establish dependencies over other organizations that cannot reciprocate resource exchanges."[153] As Richard Emerson argues, in attempts to construct a theory of power in social relations, "power implicitly resides in the other's dependency."[154] Ronald Burt agrees and states that organizations possessing those scarce resources which lead to power over other organizations can force the compliance of others (e.g. individuals, organizations, institutions).[155] For example, the literature suggests the following: consider organizations A, B and C. The exercise of organization A's power may be seen through its use of resources to gain the compliance of other organizations B and C. Thus, organization A's power can be inversely related to organization B's and C's power, and can be directly related to organization B's and C's dependence, if organizations B and C have no access to extra resources from sources other than A.[156]

This scenario very cogently describes the relational environment of grassroots NGOs in Africa, claimed by the optimist school to be working towards the development of the continent through the formulation and implementation of indigenously-defined development agendas with the assistance of foreign donors and African governments, both of which are their material, technical and financial resource bases. As such, local grassroots NGOs as well as indigenous national NGOs inevitably struggle with establishing autonomy and less dependence for their organizations, making the case that their organizational autonomy will enhance the possibilities for actually and expeditiously achieving not only "successful" substantial development, but also eventual self-reliance of their communities and the continent.[157] This study intends to determine whether interorganizational transnational network analysis, RDM and gender analysis can lend some insight and some answers to this proposition about local grassroots NGOs and national indigenous NGOs, as well as lend some direction to foreign donors and governments. Some major research questions which emanate from this are:

1 What implications does the RDM have for organizations which lack the necessary resources to meet their stated goals without the assistance of other organizations? Is autonomy possible, despite assistance?
2 Can and does networking or linking between and among organizations change the distribution of resources and hence the balance of power? Is it possible for financially and technically resourceless organizations to direct and control joint organizational endeavors? Is it possible for resourceless women's organizations to control joint organizational endeavors?
3 Are there ways to change resource dependence among organizations into facilitative cooperation, particularly when one of those organizations is a women's organization?
4 Can NGOs realistically achieve autonomy, self-directed development and eventually self-reliance while depending on foreign donors and African governments for foreign and technical assistance?

DEFINING POWER: A CONSIDERATION OF GENDER IN RDM

Application of RDM to interorganizational relations appears to have utilized the operational definitions of "power" as defined by 1) Robert Dahl: the ability of individuals to have another do what they would not otherwise do; and 2) Emerson, as previously noted: power lying asymmetrically in dependence.[158] Moreover, IOR theorists further argue that these definitions do not negate Levine and White's exchange model as much as they incorporate it.

Peter Blau, applying these definitions to IOR, specifies the conditions necessary for the possession of resources to lead to power of one organization over others:

1 an organization does not have resources to reciprocate;
2 no alternative suppliers exist;
3 no possibilities exist for using coercive power against the resource holder;
4 the organization cannot get along without the scarce resources.[159]

Blau, like other IOR scholars, has omitted one of the essential conditions for the possession of resources leading to power. That condition is to have membership in the gender group of men. Being male in a patriarchal world of power and politics lends itself the male prerogative and privilege to dominate and to have control over women's organizations in IOR through patriarchal structures. IOR has yet to deal in a serious, systematic and scholarly way with the gender issue. As an overwhelming number of NGOs, particularly from the South, are women's organizations that have become institutionalized "partners" in IOR, the gender issue cannot be sidestepped. On that basis, this study has been formulated with MYWO as the focal organization among its partners in development, requiring that gender analysis be central to this study.

APPLYING TRANSNATIONAL NETWORK ANALYSIS TO IOR PHENOMENA WITH RESOURCE DEPENDENCIES

Both Evan and Jonsson suggest similar steps to approach interorganizational studies.[160] These steps facilitate the application of the foundation of the RDM in this study. The first step is to *locate* a central or focal organization around which to center a study. The focal organization is the organization (or organizations) "that is the point of reference" which "interacts with a complement of organizations in its environment."[161] The second step is to *identify* those organizations and institutions in the focal organization's relational environment with which it interacts, bargains, coalesces or allies.

Evan refers to those organizations in the relational environment of the focal organization as "input organizations" and "output organizations" according to their functions, adapting his methodology from Talcott Parsons' general systems theory.[162] Evan defines input organizations as those organizations that provide resources to the focal organization. Output organizations are the organizations that receive the goods and services, including the organizational decisions. This overall systems analysis requires that the effects of the output organizations be traced back to both the focal organization and the input organization so as to analyze the system in its organizational entirety.[163] Figure 2.2 illustrates a general systems theory sociogram of interorganizational behavior as conceptualized by Evan. Figure 2.3 illustrates a general systems theory sociogram as applied to this study.

From another perspective, Jonsson views interorganizational relations in terms of what he calls "the game-theoretical notion of 'strategic interdependence'."[164] He defines strategic interdependence as "the ability of each participating actor [in IOR] to gain his ends [being] dependent on the behavior of other participating actors." Jonsson views the focal organization, which he refers to as the "linking pin," as the broker organization which: 1)

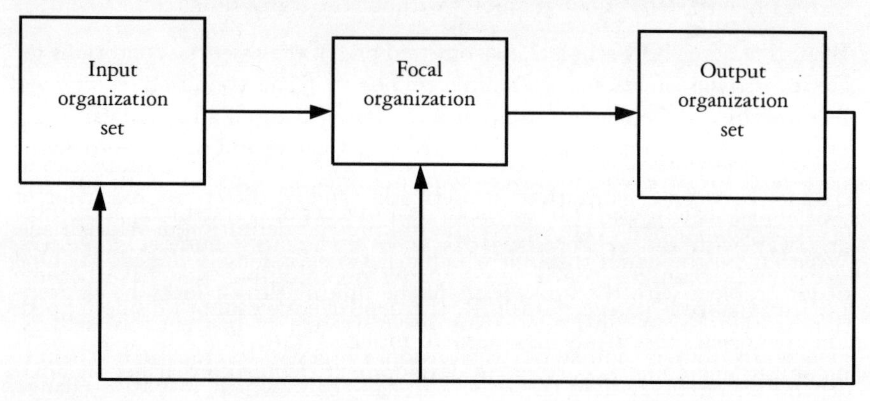

Figure 2.2 Interorganizational system
Source Evan (1976)

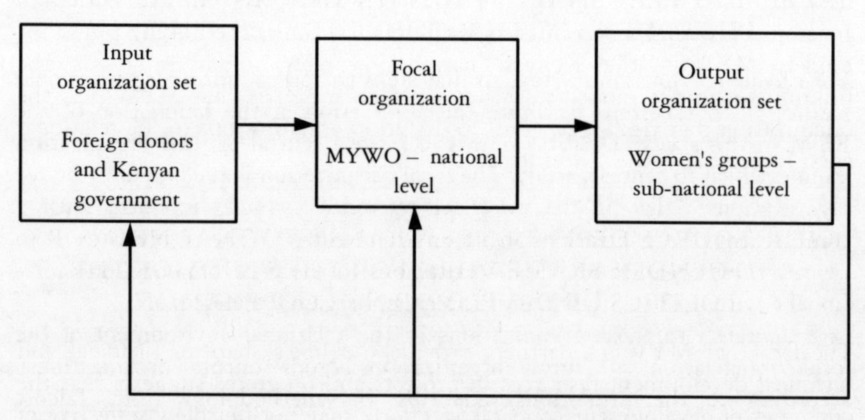

Figure 2.3 Application of general systems theory to this study

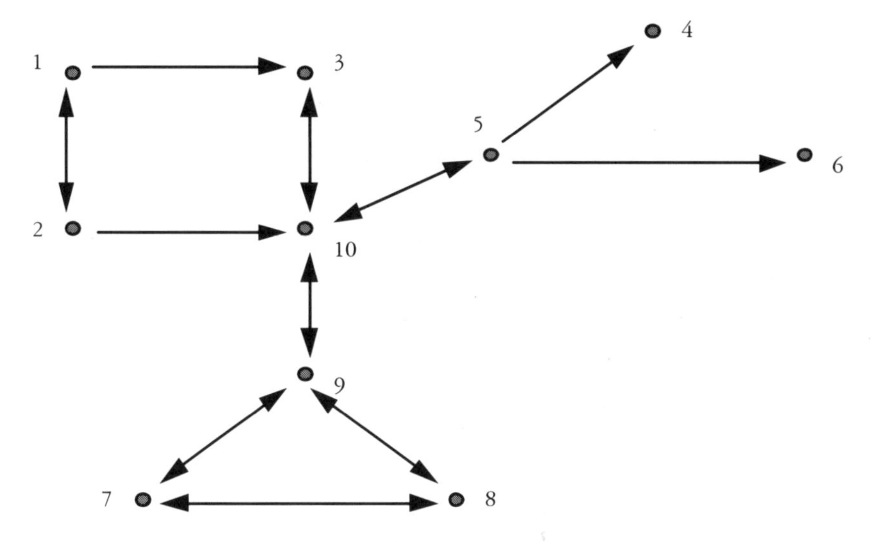

Figure 2.4 Transnational network linkages.
Source Jonsson (1986)

keeps the network of organizations together, 2) serves as the communication channel between organizations in networks, and 3) directs the behavior of other organizations.[165] He adopts the linking pin definition of Aldrich and Whetten as "the nodes through which a network is loosely joined."[166] One major problem with the broker role of the linking pin in Jonsson's description is that the organization assuming this role is not entirely selfless. Hence, its role is not static as Jonsson suggests. Its role may change, depending on its individual organizational goals (which may also change over time), its self-interest and its relative power (and dependence) vis-à-vis other organizations in the network. Figure 2.4 illustrates hypothetical network links with a linking pin as conceptualized by Jonsson. Although Jonsson conceptualizes an international organization as the linking pin in his model, MYWO—the Kenyan national women's organization—is the linking pin in the model of this study. The reason is that MYWO is the organization that ties together all of the other organizations in the relational environment. It is the center for all activities.

THE RESOURCE DEPENDENCE, TRANSNATIONAL NETWORK AND GENDER FRAMEWORK FOR THE STUDY OF THE POLITICS OF DEVELOPMENT COOPERATION

The fact that MYWO's partners provide the lion's share of financial and technical development resources to projects and programs suggests that the outcomes of development cooperative efforts may not be solely reflective of MYWO's defined goals and priorities, but may instead reflect unequal power and gender relations, and hence the goals and priorities of its partners—foreign donors and the Kenyan government. The degree to which MYWO has been able to direct development as the autonomous body it claimed to be at certain periods in its history, free from the pressures of the exchanges made with foreign donors and the Kenyan government, determines its institutional effectiveness as a development NGO, in spite of dependencies inherent in unequal resource contributions and unequal gender relations in development partnerships. The degree to which development partnerships have been able to meet MYWO's development goals, despite the unequal distribution of power within the partnerships, will hopefully give some indication as to how development successes might be optimized in the future, using available resources in the organizational relational environment. Analyzing the politics of gender in IOR cooperation is extremely important, in that gender differences reinforce and reflect the politics of unequal resources as well as shape the current character of development partnerships in contemporary Africa.

This chapter has merged four distinct areas of research: 1) IOR theory; 2) NGOs in Africa; 3) development theory, and 4) women in IOR. It has also reviewed the relevant literature so as to construct an appropriate framework for analyzing the research problem at hand—the politics of development cooperation. The next chapter will provide a political history of MYWO, as the necessary background information for the presentation and discussion of the fieldwork in Kenya.

3

THE EVOLUTION OF MYWO FROM 1952 TO 1992

This chapter will present a political history of Maendeleo Ya Wanawake (MYWO) in Kenya. It will discuss the evolution of the organization from 1952 to the present, noting the establishment of and changes to its interorganizational relationships with donors in its relational environment—namely, the colonial and post-colonial Kenyan governments, and foreign and international governmental and non-governmental organizations and businesses.

This chapter is divided into two parts. Part I addresses the period from 1952 until 1987, during which time MYWO moved full circle, existing officially as a colonial state organization between 1952 and 1961, and again becoming a post-colonial state organization in 1987. Part 2 deals with the period from 1987 until 1992, during which time MYWO frenetically attempted to reconfigure itself among its interorganizational partners—foreign donors and the Kenyan government. Special attention will be paid to 1) the 1989 KMYWO elections and the subsequent changes in the organization's interorganizational relationships; and 2) the 1991–2 period, when MYWO and its partners were involved in heated squabbles over multipartyism. The central argument of this chapter revolves around the question of whether or not MYWO could ever really be considered a genuine and autonomous NGO, truly committed to representing African women at the grassroots and their interests. Providing contextual historical information is critical as this chapter traces the linkages between MYWO and its interorganizational partners in a transnational network.

PART I: MYWO AND THE STATE OF EMERGENCY

MYWO has a very complex and seemingly contradictory history. Its roots were planted during the colonial period from which, its critics argue, it has not been able to disentangle itself. Its proponents, however, have chosen to forget its beginnings, or simply not to acknowledge them, particularly after the organization assumed a more nationalistic posture in the 1960s.

MYWO was officially founded in 1952 by European colonial women who were the wives and relatives of British colonial officers. Its founding occurred simultaneously with the declaration of the State of Emergency in Kenya, when virtually all African political organizations were outlawed and African leaders detained. MYWO was officially recognized by the colonial state at this time because of the critical role it was to play in repressing Kenyan nationalism.

Audrey Wipper, who completed the most extensive study of MYWO from 1952 to the mid-1970s, has argued that it was mere coincidence that the organization was founded at the same time that the Emergency was declared. She maintains that MYWO played no covert conspiratorial role with the colonial state in attempting to subvert the Kenyan national resistance movement.[1] Instead, she argues, it was the European women's empathy for the plight of African women and their aversion to African women's lag behind African men in development, along with their spirit of *noblesse oblige*, that were the major forces for their creation of MYWO.[2]

Cora Presley, who has written on Kikuyu women and Mau Mau, provides evidence that suggests quite the opposite—that it was not mere coincidence that MYWO was officially founded at the height of the State of Emergency. Without entering the debate as to why MYWO was created, Presley demonstrates that MYWO was indeed used by the colonial government to subvert Mau Mau by gathering information about resistance and guerrilla activities. She argues that MYWO's aim was to persuade Mau Mau adherents to abandon the movement.[3] Presley recounts how African women were told to abandon their commitment to Mau Mau or lose the "humanitarian" services MYWO clubs provided.[4] Archival data which will be presented in the following pages supports Presley's argument and turns Wipper's assertion on its head.

MYWO was modeled after English women's tea clubs. The first president was Miss Nancy Shepherd, the granddaughter of one of Kenya's first missionaries, the Venerable Archdeacon H.K. Binns.[5] Shepherd became the principal of the Jeanes School (now the Kenya Institute of Administration) and the colonial supervisor of MYWO's activities. This was her domain as Assistant Minister for Women and Girls and later as Assistant Minister of Community Development and Rehabilitation in the Department of Community Development of the colonial government. Under Shepherd's direction, the Jeanes School became the training center for MYWO African women leaders and the springboard for MYWO clubs.[6] The Department of Community Development of the colonial government had realized early on the need to train African women who could "accelerate Maendeleo's expansion in the villages where most potential members resided."[7] Its aim was to create MYWO local clubs of African women members to carry out anti-nationalist activities.

Although African "woman power" had been declared necessary for the effective administration of the colony, no Kenyan woman was allowed to be in the effective leadership of MYWO. The first headquarters committee of MYWO was composed entirely of European women, apart from the *honorary* vice chair, Harriet Musoke, a Ugandan woman who was then a probation officer in the colonial government.[8]

> The original constitution of MYWO stipulated that the organization was formed to develop and improve conditions for Africans through social intercourse by bringing the women together, encouraging neighborliness and cooperation and education which could be largely informal and practical in scope.[9]

The curriculum for the first class at the Jeanes School focused on sewing, cooking, child welfare, hygiene, games, singing, dancing and outdoor games.[10] It is reported that not until later did the curriculum embrace areas such as community development which had been, since colonial intrusion, the domain of Europeans.[11]

Although the organization gained membership of African women, with a reported 172 clubs in 1952 and a total of 72 women leaders passing through Homecraft at the Jeanes School the same year,[12] there was resistance to MYWO. From one direction, colonial settlers, in what were then called the "White Highlands" of Kenya, demonstrated extreme resistance, as they opposed any education for Africans. Because many African women were squatters on or near colonial farms, the colonialists felt that the "ignorance of the African woman was a great advantage in terms of having an available supply of cheap labor."[13] From another direction, there was resistance from the African population, especially in Central Province, mostly among the Kikuyu but also among the Meru and Embu. They felt that the colonial authorities were using MYWO as a tool to suppress the efforts of Mau Mau. They believed that,

> in Central Province, where most [white] club leaders were wives of white DOs [colonial district officers], African club members would give information on the latest movements of their husbands day and night. This way, those whose movements were deemed suspicious would be picked up and locked up in concentration camps as members of Mau Mau.[14]

African women might have given this information, as many of them were not suspicious of Shepherd's motives. She is still very much revered as a "very devoted social worker who mixed freely with Africans."[15]

There is ample evidence in the British colonial archives to support the allegation that MYWO was, in fact, an informant organization and part and parcel of the colonial police state. After the reported "outbreak" of Mau Mau in October 1952, the colonial government began a systematic coordinated program code-named "Anvil" designed to "clean up Nairobi, disrupt Mau Mau organizations and stop Mau Mau progress."[16] One of the main methods used to accomplish this was the suppression of African women. Although African and Africanist historians have traditionally recognized and acknowledged that the top Mau Mau leaders were men, namely Dedan Kimathi, Stanley Mathenge, General China and Kimbo,[17] women have become belatedly recognized as being "among the most fanatical supporters of this secret society."[18] Women's active resistance roles in Mau Mau have been neglected by most scholars, with a few exceptions including recent works by Tabitha Kanogo and Presley, who both write about Mau Mau and women. Archival data indicate that British colonial authorities reported that it was in fact women who moved between the settled areas and the forests from where Mau Mau launched their attacks. Colonial reports indicate that "They [the women] carry food and information, and when necessary, take part in the fighting. They also help to intimidate the peasants and conduct the oath-taking ceremonies."[19] The Provincial Commissioner of Central Province was categorical about the role of women:

> In information to all sections the women present a particular problem and one in which special attention must be paid, since it appears that it is the women, as much as if not more than the men, who are keeping the spirit of Mau Mau alive.[20]

In a colonial newsletter T.G. Askwith, the Commissioner for Community Development, blatantly revealed the government's intended use of women's groups to counter subversive elements.[21] He stated: "African women's clubs, particularly those in Emergency areas, are regarded by the Kenya Government as valuable rallying points which are doing a considerable amount to overcome the influence of Mau Mau."[22]

In exchange for efforts to overcome Mau Mau, Askwith revealed at a conference arranged by the Ministry for Community Development and Rehabilitation that the government had been allocating "development funds for the women's clubs at the rate of £12,000 sterling a year."[23] Therefore, the colonial state had begun government financial support for MYWO. The government had also begun the practice of exempting compliant Kenyan women from the compulsory five to six days of forced labor. In effect, the colonial government's financial assistance and exemption from forced labor were clandestine exchanges and rewards to entice African women to work against Mau Mau.[24] These exchanges appear to be the first interorganizational relations, in cash and in kind, in which MYWO and the colonial government would engage in their interorganizational networks.

Other colonial officials and British aristocracy also used women's clubs to attempt to crush Mau Mau. In a conference in 1955 in Kiambu, organized by MYWO Homecraft Officer Mrs Winifred Moore, District Commissioner Frank Lloyd opened the conference by saying that it was up to women to fight Mau Mau influence and help maintain law and order.[25] Lady Mary Baring, wife of Kenya's governor, and her mother, Lady Grey, also traveled and appealed to women's groups. Lady Baring, in a speech to the women, praised the work of women's groups and told them that "Kikuyu women could do a lot to help finish Mau Mau trouble by working well together, sharing their pleasure and being good friends to all."[26] In response, the women's club leader, Louise Williams, told Lady Baring that "the club agreed that the Mau Mau troubles hindered their programmes but they were determined to help in ending Mau Mau and use their club activities to improve living conditions in the village."[27] Colonial officials in Kiambu further endorsed the idea by demanding more women's clubs in the area.[28]

Efforts to contain Mau Mau were also promoted outside Kikuyu areas in Central Province. These efforts suggest that the movement may have been more widespread than the Kikuyu areas, as conventional knowledge suggests. For example, in 1954 Lady Baring addressed women's groups in Machakos, Eastern Province. Among them was one of the oldest groups in the area, Iveti's Women's Club. Mrs Penwill, a MYWO women's group leader, identified this group as having raised the alarm about a "gang" in the area. This "gang" was later wiped out by the colonial created and armed Kamba Home Guard.[29] Mrs Mary David, another MYWO women's group leader in this area during Lady Baring's visit, expressed gratitude to the colonial government on the part of all Kamba women (Eastern Province) that Mau Mau had not come into the district. The proliferation of groups in this area further demonstrated the extent to which the campaign to end Mau Mau had penetrated. From 45 in 1953, the number of women's groups in the Machakos area had doubled to 90 by 1954.[30]

Clubs in other areas had been created as well. The *Community Development Annual Report for 1952* noted the type of club work that had been done in the following areas:

Area	*Type of club work*
Nyeri	Had some of the most successful women leaders
Fort Hall	Had 6 clubs, numbers declined with Emergency
Kiambu	Had 33 more to open at Githunguri
Embu	Successful club at Boma
Meru	Plans for future
Garissa	1 Somali woman did a little with DC's wife
Kitui	Some work but no evaluation
South Nyanza	19 clubs
Baringo	Several established clubs
Tambach	Improved with European women
Nakuru	2 excellent clubs
Machakos	Over 30 clubs
Mombasa	2 trainees to work with women of all ethnic groups
North Nyanza	In the forefront of all districts. Women's Institutes are the local names of MYWO

Central Nyanza	12 groups
Nairobi	Started its first group at Kaloleni. Other groups expected
Kilifi	Trainees went to Embu for further training
Taita Taveta	Just received qualified teachers[31]

After 1952 the mobilization of MYWO clubs continued, led by European women and African women graduates of the Jeanes School. For example, in 1954 Mrs Alice arap Kirui established a Homecraft Center for Kipsigis women in Kericho, Rift Valley Province, in which 35 women enrolled. In that same year a group of European women, including Shepherd, built a club for African women in Karen to teach homecraft and domestic science.[32] In Kitale (1955) and Nakuru (1954), both Rift Valley Province, courses were also offered (for a fee) in sewing and knitting, and domestic science respectively. The Kitale group was being assisted by Mrs Gladys Ombara, an ex-Jeanes School student and by a government grant of £25.[33] This is another instance on record of an MYWO group receiving financial assistance from the colonial government. In Nakuru, the colonial mayoress for the area, W.H. Sayer, assisted a group and told the women that as a follow-up to the course she would check the "cleanliness" of their homes.[34]

By 1955, there were over 400 African women's clubs in Kenya with a membership of over 40,000; and, as the Colonial Press Office reported, "all of these clubs were part of the country-wide Maendeleo Ya Wanawake Movement sponsored by the Government."[35] To further this effort, more resources were poured into the campaign against Mau Mau. For example, in Kenya Newsletter No. 90, it was stated that African women's clubs sponsored by the Ministry of Community Development's MYWO movement had been allocated £40,500 sterling for the purchase of equipment, furniture and materials during the three and a half years covered by Kenya's new Development Plan.[36] Concomitantly, the Council of Ministers of the colonial government argued that continued supervision of women's clubs was critical to keep them from Mau Mau influence.[37] Based on this, the colonial administration decided to continue channeling funds to the groups even after overall cuts had been made in the Emergency budget in Kenya in 1958. It was believed that if the funds were withdrawn, clubs would fail and British efforts to destroy Mau Mau would be compromised.[38] Thus, the colonial government would again provide financial assistance to MYWO in their interorganizational relationship. Its aims were to 1) keep MYWO clubs beholden; and 2) keep the colonial state cushioned and secure from Mau Mau resistance.

MYWO had also begun establishing relations with foreign donors who were providing financial assistance to the organization. For example, as early as 1955, the Associated Country Women of the World (ACWW), an INGO, provided financial and technical assistance to MYWO.[39] In 1959–60, the United Nations International Children's Emergency Fund (UNICEF) additionally donated a grant of US $88,500 (which the colonial government matched) for the expansion of health and homecraft activities at the village level.[40]

The creation and financing of African women's clubs represent only a fraction of the colonial state's efforts to crush African resistance to British colonialism. Working in conjunction with the state-assisted African women's club movement was a program of psychological warfare that was declared by the colonial government against Kenyans in 1953. This psychological warfare was a mass propaganda campaign directed towards various "classes" of people defined by the colonial government. The intent of this campaign was to annihilate Mau Mau totally. One of the main avenues that the government used to filter its propaganda was through women's groups.[41] Colonial records indicate that the government was "planning to start a fortnightly women's paper in Kikuyu to meet the need for the Emergency and political news for women."[42] By 1954, the paper was not only published in Kikuyu, but in Kikamba, Luo and Kiswahili.[43] The groups on which the government focused primarily were the Kikuyu, Embu and Meru as they were considered to be the bulk of the Mau Mau "terrorist" strength."[44]

There is conflicting evidence about the extent of the circulation of the paper. In 1952, records indicate that the government produced and distributed a quarter of a million copies per week. In 1954, however, it was reported that the paper had a circulation of 18,000.[45] In addition to the paper, the government used radio, posters, captions, films, cinemas and vans for distributing its propaganda.[46]

The "themes" of the colonial government's psychological warfare campaign were:

1 Mau Mau was an evil thing bringing hardship and misery to thousands, and the task of destroying this thing lay with Africans themselves. Europeans and Asians would help;
2 The white settlement was critical to the economy;
3 Within the government there was the possibility of advancement for all, so that all might march to prosperity together;
4 "Prestige" was due to African government servants, particularly chiefs, giving Africans a sense of loyalty to the Commonwealth.[47]

The colonial government focused on brainwashing the women with this psychological warfare propaganda. By the systematic propagation of this "very hotted up information"[48] disguised as women's political news, the government further relied on women to filter this information into the community through MYWO groups. The government's position was that: "We shall give courses to picked leaders at the Jeanes School (this has been done before) and then send them out to proselytize in the Reserve."[49]

Africanization of MYWO and disassociation from the colonial state: revolution from within

While the colonial government expanded MYWO's women's clubs and secretly plotted psychological warfare, Kenyan women were also developing a clandestine resistance agenda of their own. From inside MYWO, Kenyan women who had been trained at the Jeanes School to take lower-level domestic positions within MYWO orchestrated a coup against the colonial leadership of the organization. In 1961, at the Annual General Meeting of MYWO, the British leadership of the organization was voted out and a decision was made to change the policies of the MYWO movement for the benefit of its majority African members.[50] Mrs Phoebe Asiyo displaced Shepherd, becoming MYWO's first African President.

Asiyo's words indicate the inevitability of this revolution from within the organization:

> When I joined Maendeleo in 1956, it was after fruitless efforts to set up our own indigenous African controlled national women's organization. We failed because of lack of womanpower and financial problems. Still when we joined Maendeleo, our clear intention was to take over its leadership gradually because as African women, we understood our problems better. Things like queen cake baking and other European dishes which provided the bulk of Maendeleo clubs' syllabus were not the answers to our problems. African children still had less than enough to eat. True, the white ladies tried hard and they were sympathetic but they did not understand the African woman's needs.[51]

Thus, it was declared that MYWO be converted into a more meaningful indigenous African woman's organization, not only concerned with "feminine-domestic" issues, but more importantly with the overall welfare of Kenyans.

There was much surprise and resistance to the leadership takeover of MYWO, namely by Shepherd, European settler wives and colonial government officers. Shepherd was devastated that she would no longer be in control of the organization; and the settler wives found it difficult to accept Asiyo as their leader and treat her with respect and equality. Asiyo noted, "The settler wives had this peculiar idea that since I was African I should be served tea in the kitchen."[52] In addition, the colonial government, whose plans of infiltration and psychological warfare had taken an unexpected turn, were baffled. They had tried unsuccessfully to hold on to MYWO. The new African women leadership reacted in this way:

> When the first African dominated National Executive Committee of Maendeleo in its first session decided to break away from the central colonial government in favour of autonomy, we met with a lot of oppo-

> sition. This patriotic move on our part was condemned in public and the colonial establishment even went out of its way to use brainwashed African women to attack us.[53]

When the colonial government realized that they could not hold the independent spirit of these African MYWO women under their thumb, they kicked them out of the government-owned premises, with only one old typewriter and one file with a copy of the MYWO constitution, one desk, one wooden chair and a grant of Kenya £20,000 (US $56,000).[54] What the colonialists were forced to face was that a critical pillar of their colonial apparatus had been dislodged. African women had disassociated MYWO from the colonial state and, in effect, declared MYWO an autonomous non-governmental organization (NGO). One might argue that the grant of Kenya £20,000 was an attempt to maintain some interorganizational linkages and have some influence on MYWO's agenda.

Within two years of the takeover of the organization, the new MYWO had run out of finances. The colonial government grant had been spent. The Africanized MYWO did not fold, however. Instead, it continued the implementation of three programs it deemed more "suitable" for the African woman.[55] Its new leaders agreed to work on a strictly voluntary basis, without the rewards and exchanges from the colonial government as before. They used their personal finances and monies that they had amassed by sponsoring fundraisers. In doing these things, MYWO effectively established its autonomy from the colonial state. Moreover, MYWO declared that its organizational autonomy was more important than interorganizational exchanges with the colonial state.

During this period of transition to an African autonomous organization, MYWO survived principally because of: 1) the determination of the new African women leaders in the organization; 2) the assistance of African nationalist and pan-Africanist leaders such as (a) Tom Mboya, a Kenyan nationalist who would later become the Secretary General of KANU and the Minister of Economic Planning in the first post-colonial Kenyan government; and (b) Kwame Nkrumah, the first president of post-colonial Ghana and one of the pioneers of pan-Africanism; and 3) foreign donors who would provide financial and technical assistance to the organization.

Some of the first foreign donors to assist the autonomous MYWO were the League of American Women Voters, ACWW and UNICEF.[56] ACWW and UNICEF basically continued the interorganizational relationships that had been established with the colonial MYWO. Noting that these relationships were created at this time is critical in that this is the first time MYWO—the autonomous NGO—established interorganizational linkages with foreign donors. This collaboration with foreign and government donors would continue and expand to include the ministries of the 1963 post-colonial Kenyan government, even after Asiyo's resignation as chair of MYWO that year. It is no surprise that the autonomous Africanized MYWO survived, as Kenyan women had already proved their strength as warriors in adverse times.

From nationalist politics (1961–3) to grassroots mobilization, government collaboration and self-definition (1963–71)

Despite MYWO's autonomous status and its focus on social welfare, during the period between Africanization and the wake of Kenya's liberation it became imperative that MYWO formally and publicly involved itself in the independence struggle. In a nationalist undertaking, Asiyo led MYWO in a petition to the colonial government against the detention of Mzee Jomo Kenyatta and his colleagues.[57] In another nationalist undertaking, Asiyo and an MYWO delegation visited Kenyatta in prison. They recounted:

> We (MYWO) stood for the release of Jomo Kenyatta and the true independence of Kenya. At Lodwar, we gave Mzee flowers and vegetables and in appreciation he remarked that women saw deeper than men in all spheres. He promised us a place in his government and this was reflected in the original KANU manifesto which said among other things that it would work side by side with Maendeleo Ya Wanawake.[58]

Whether it was desired or not, MYWO had, by Kenyatta's promise, become inextricably tied to the impending first post-colonial government. After all, it was African women in Mau Mau and in MYWO who had demonstrated that revolution from within was indeed possible. However, in an attempt either to wrench itself from KANU and politics completely or to avoid killing itself by siding with what might have become a losing party, or perhaps even to maximize its negotiating power with the new government which would replace the colonial government, MYWO under Asiyo's leadership refused to commit itself in support of either KANU or KADU (Kenya African Democratic Union)—the two Kenyan opposition independence parties.[59]

With KANU having won the elections in 1963, and more importantly with the colonial regime having been ousted and the first post-colonial government formed, MYWO felt that it had also won. The women felt that they were finally rid of the repressive regime which had intended to make Kenya a "white man's country,"[60] which had jailed and killed Kenyan leaders and which had used unsuspecting African women to conspire with them. In expressing MYWO's new role, Asiyo noted that,

> Immediately after independence we were frequently invited to sit in important policy making meetings and the new majority government

> consulted very closely with women not only on social service matters but also on other matters of national importance. That is how Maendeleo got its current format of an autonomous national women's organization.[61]

How stable and effective this new role was, and how autonomous the NGO MYWO would actually be, would be a frequently recurring question for MYWO, the Kenyan state and scholars.

Despite the optimistic outlook with which Asiyo approached the era of independence, after independence MYWO did not prove to be a top priority of the government. From 1963 to 1976, the new post-colonial government did not recognize and cultivate the potential of women.[62] Certainly, the post-colonial government did not have the same degree of financial and administrative resources to allocate to Kenyan women that the colonial government had had to allocate to European women.

Maria Nzomo, noted scholar of Kenyan politics, provides a very poignant analysis of the relationship of women to the newly independent government. She states, "During the 1960s and 1970s, the government completely ignored women's issues. It acted as if the gender question did not exist in Kenya."[63] Her argument suggests that the government's non-prioritization of women goes beyond the possession of sufficient resources. Fundamentally, she implies, the government's choices involved values, priorities, power and patriarchy. Her argument is solidly supported by the manner in which Kenyatta formed his regime and government, and reconstituted the state. Kenyatta concentrated power at the top in patriarchal institutions, particularly 1) the Gikuyu Embu Meru Association (GEMA), the political machine that supported the Kenyatta regime and maintained Kikuyu hegemonic control in Kenya,[64] and 2) KANU. Kenyatta also surrounded himself with an all-male vanguard. Moreover, his government and state policies were much more male and urban-focused rather than rural and female-focused.[65] All of these factors had an impact on the sustenance of MYWO groups and their expansion in the rural areas.

There is conflicting evidence as to MYWO's effectiveness as a mobilizer of women's groups after 1963. At the thirtieth year anniversary celebration of the organization, it was stated that MYWO's effect on the mobilization of women became insignificant immediately following independence, especially in the rural areas. The organization's lack of mobilization power was a direct consequence of the destruction of the colonial administrative network from the capital, Nairobi, to the Local Native Councils in the rural areas. MYWO's lack of mobilization effectiveness was further exacerbated by: 1) the many African women who refused to be part of the organization because of its colonial legacy, particularly in the Central Province; and 2) the Kenyatta government's overall lack of concern for women's groups and their activities. Ironically, it was the colonial government that had decentralized

MYWO and taken it to the grassroots, whereas the first post-colonial government would break these links.

The second African chair of MYWO, Mrs Jael Mbogo, who assumed the leadership of MYWO in 1963 after Asiyo's resignation, poses the most conflicting and challenging evidence to MYWO's decreasing effectiveness as an organizer of women's groups. Mbogo claimed that the membership of the organization increased phenomenally during her tenure as chair, while other records indicate that MYWO membership had waned. For example, Mbogo, who had spent much of her time organizing MYWO groups in the Rift Valley, says: "During my term in office as President (1963–7), Maendeleo clubs rose from a mere 4,000 to 85,000 with each claiming a membership of over 3,000 women. I also started 3,500 nursery schools run by MYWO."[66]

Mbogo further indicated that during this time MYWO received government financial assistance from Dr Gikonyo Kiano, the then Minister of Commerce, for a handicraft exhibition and for international exhibitions to explore world markets.[67] This financial assistance is important because it illustrates the furthering of interorganizational linkages between MYWO and the post-colonial Kenyan government in development cooperation.

Mbogo's successor, Mrs Elizabeth Mwenda, who was chair of MYWO from 1967 to 1968, is recognized by MYWO as having brought 50,000 more women members into the organization, and as having created 16,500 more nursery schools.[68] During her administration, the government further cooperated with MYWO by donating vehicles for their work. With regard to this interorganizational donation, the then Director of Community Development, Mr Jonathan Njenga stipulated, "Depending on your ability to run the new vehicle in addition to the VW you already have, we shall consider your request to have more vehicles to cover the provinces."[69]

Mrs Ruth Habwe, the chair of MYWO 1968–71, seems to have followed the path of her predecessor Mwenda by creating more women's groups. Habwe is credited with expanding MYWO to Lamu Island where she established its first branch.[70] In addition, during her term in office MYWO began to function as a political pressure group. Records of MYWO's Annual General Meeting in 1968 are demonstrative. At this meeting, MYWO passed several resolutions concerning the rights and demands of women. One of the resolutions called for equal employment terms with men in the public and private sector. Another resolution was for an increase in the number of places for women students at the University of Nairobi.[71]

There are very scanty records on MYWO groups during these periods. In explaining why records have not have been kept, a national officer of MYWO from 1971 to 1980, stated, "One reason there is not a lot of information on file is that women were doers, out in the field."[72] Records do indicate, however, that women's groups did spring up during this period independently of MYWO. Many of these groups were comprised of women who provided each other with mutual assistance to meet needs of their community, lessen their individual workloads and save money. One example is the *mabati* (roofing) women's groups, who thatched roofs.[73]

National and international growth of MYWO under Jane Kiano (1971–84)

Particular attention must be paid to the chairship of Jane Kiano, for she has emerged as one of the most important and dynamic leaders of MYWO. Her name reverberated in the interviews with women throughout Kenya, in their discussions of MYWO "before the affiliation to KANU."

Kiano was the chair of MYWO for 13 years, from 1971 to 1984. She held the longest running term in the history of the organization. Today, she is one of the best-known women leaders in Kenya and the patron of MYWO. She is also criticized by some outside the organization, particularly intellectuals and leaders of other women's organizations, who argue that she has compromised the potential both of women in Kenya and of a more radical women's movement.[74]

Kiano holds to her credit the "unprecedented growth" of MYWO after its decline in the 1960s and 1970s. From a reported 2,085 MYWO and non-MYWO women's groups existing in 35 of the 40 districts in Kenya between 1971 and 1975, it is reported that "Mrs Kiano saw MYWO (alone) become 6,000 affiliated women's groups with, by 1983, an individual membership of 327,000."[75] Still, this figure hardly sounds astounding given the reported figures of the Mbogo administration which claimed a membership of 85,000 affiliated groups.

Kiano is the wife of Dr Gikonyo Kiano, the current chairman of the Kenya Broadcasting Corporation (KBC) and the Muranga district KANU branch. He was also the Minister of Commerce during the first post-colonial government. Mrs Kiano was elected chair of MYWO in one of the "toughest contests" in the history of the organization.[76] Kiano defeated Mrs Nyiva Mwenda, a former KANU Member of Parliament (MP) for Kitui Central and the wife of a former chief justice of Kenya, Mr Kitili Mwenda. After Mrs Kiano's first victory, she was re-elected chair of MYWO several times with no significant opposition. As detailed hereafter, her marriage to Dr Kiano had a direct influence on her victories.

MYWO enjoyed relative successes during Mrs Kiano's terms in office. Kiano also enjoyed organizational support from the state, which had been unprecedented since colonial times. For example, during Kiano's term, the MYWO headquarters—Maendeleo House, a project which cost Ksh 14.5 million (almost US $1,600,000)—was completed.[77] In order to accomplish this, MYWO had not only the support of its members, but also "the strong support of the government of Kenya, local banks, voluntary organizations and international bodies."[78] In addition, the Maendeleo Handicraft Shop was converted into a successful cooperative and business inspired by Dr Kiano, who assisted the organization through his ministry. Mrs Kiano also personally secured major funding from the Konrad Adenauer Foundation for MYWO's Leadership Development (LD) program. In addition, during her administration, long-term funding for MYWO's three other major programs—Maternal Child Health/Family Planning (MCH/FP), Special Energy Program–Jiko (SEP–Jiko) and Nutrition were secured.

There were several factors which led to Kiano's successes. They were: 1) her personal style of leadership; 2) MYWO's homemaking ideology; 3) her diverse economic and political alliances which enhanced MYWO's ability to deliver goods and services; 4) MYWO's endorsement and participation in *harambees* (see discussion following); and 5) her high profile in women's affairs internationally.

Kiano had a very strong personalized leadership style. She made personal requests to individual women leaders for their specific and organizational assistance. Very early on, Kiano began to create a network for assistance through her involvement with women's groups and development projects. She has stated,

> Even before I became involved in the leadership of Maendeleo Ya Wanawake I was involved and I am still involved in community work in Muranga (Central Province), leading women in self help projects such as the building of maternity clinics, nursery schools, water projects, adult education classes and in mabati groups.[79]

Kiano also used her personal leadership style to request individual men connected to the state in official capacities to assist the organization. One of Kiano's close counterparts gave this example: "Men used to help by support and anonymous donations, because MYWO was a strong organization."[80] She stated that these were men with power and money, involved in the government of Kenya. Kiano's husband ranked prominently among them: "Dr Julius Kiano used to boost morale and guide MYWO."[81]

Mrs Kiano even used her personal relationship with Kenyatta to raise MYWO issues. She is a Kikuyu from Central Province and as such, because of Kenyatta's "ethnic politics," she was guaranteed an audience with him.[82] In addition, her husband was part and parcel of the ruling machinery. He was a cabinet member, a close personal and political ally of Kenyatta, and a key member and officer of the GEMA.[83] Stated differently, Dr Kiano was appropriately placed in the government to lobby on his wife's behalf, and more importantly to direct monies to MYWO as the government's Minister of Commerce. In important ways, MYWO was dependent on Mrs Kiano's personal networks.

MYWO was also successful because of its conservative ideology, which did not challenge Kenya's patriarchy. Kiano was able to meet and work with Kenyatta because the organization did not make radical demands for women's rights.[84] Instead, MYWO concentrated on non-threatening feminine "homemaking' issues, at least most of the time. When MYWO did tread on issues that might be construed as "non-feminine" such as housing allowances for women, it was very careful to approach Kenyatta in a non-aggressive, unchallenging, soft, mild and stereotypically (and often deceptively) subservient ladylike manner. Kiano's strategy in using this approach was to avoid publicly disturbing the gendered spheres of power. The way this played out in interorganizational relations with the government was that whatever MYWO requested, Kiano somehow, directly or indirectly, linked to "homemaking." She was careful not to bruise male egos.

Fundamentally, MYWO was a welfare and development organization under Kiano. It gained attention and respect as a "women's" organization from the state because of its apolitical, subtle and stereotypical feminine approach, as opposed to a political, up-front, more radical feminist approach. Kiano publicly stood against women's liberation and equality, not really fighting for women's rights directly. Ironically, she was and still is the epitome of a liberated woman and a feminist. Despite this, the role that she created for MYWO was one couched in dependent, subservient politics in its interorganizational relations with the Kenyan government.

MYWO was further successful because Kiano used non-exclusionary alliance-building tactics to create coalitions that were non-ethnic and non-racial. Kiano worked with all women—African, European, Asian, citizen, expatriate, rural, urban—although Europeans and Asians were fewer than Africans in the organization's membership.

Shepherd and a number of her European cohorts, for example, continued to work with MYWO during Kiano's administration. There were also some Asian members, but much of the Asian community had left at Kenya's independence.[85] The Asian women who remained in Kenya had limited involvement in MYWO, with the exception of at least one high-profile Indian woman in the organization, Mrs Dar. She was the national treasurer under Kiano, from 1971 to 1980. Dar used her influence in the Asian community to secure economic assistance for MYWO from the local capitalist class, who were Indian. Some of them provided assistance, albeit quietly, wanting to remain anonymous. One example of a consistent, dependable Indian supporter was Julie Manji of the House of Manji, to whom Dar went for financial support for the organization. Manji was able to provide contributions and assistance because the House of Manji, her father's business, was and still is one of the wealthiest and most prosperous local industries in Kenya.[86] The House of Manji became a key interorganizational financial supporter of MYWO in the 1970s, though anonymously.

It is important to point out that this interorganizational support from Manji was not for MYWO's programs or priorities necessarily; rather it was given because individual women in MYWO's leadership had personal ties to Manji. Julie Manji, Dar, and Kiano were all from the same district area in

Central Province—Nyeri. Moreover, both MYWO and the House of Manji stood to benefit from Manji's financial contributions to the organization. That is, Manji's business could benefit in that Indian industries in the 1970s needed state sanctioning to remain in business in Kenya, and Mrs Kiano was well connected to the state, being herself an extension of it. In exchange for financial assistance, she could lobby her husband, the government, even the head of state, Kenyatta, on the House of Manji's behalf.

Kiano's alliance building extended to include not only the government of Kenya but also foreign donors. With the government, the marriage between the Kianos informally made a strategic alliance between MYWO and the ministries in which Gikonyo, Kiano's husband, was very well placed to issue assistance to MYWO. Unofficially, Mrs Kiano also made alliances with the Ministry of Culture and Social Services (MCSS) who provided an MCSS officer and the use of a land vehicle for MYWO's development trips into the countryside. Occasionally, MCSS also provided grants for MYWO projects, hence establishing more interorganizational linkages between MYWO and the government.[87] In 1984, however, this relationship between MYWO and MCSS nearly came to a head when Dr Kiano and the MCSS minister, Kenneth Matiba, challenged each other in an election in Muranga. Matiba won the election and Kiano charged that Matiba had bought the votes of the constituency.[88] Friction thus surfaced between MYWO and the MCSS, thereby causing strains in their interorganizational relations.

With foreign donors, Kiano also made short-term and long-term alliances during her term as chair of MYWO. Some of these donors included the Kenya Lions Club and Coca-Cola–Kenya. Kiano used these alliance to deliver goods and services to persons in need. Much of this, though not all, went to women.[89] For example, the Lions Club, in a short-term interorganizational relationship with MYWO, donated wheelchairs to disabled persons in the North Eastern Province. Coca-Cola, in a longer-term interorganizational relationship with MYWO, set up kiosks for MYWO members in the Eastleigh area of Nairobi so that some women might generate income for themselves.[90] Kiano also maintained relations with the ACWW. They continued their donor support for MYWO by providing, among other things, the salary of a field worker and a vehicle for MYWO to be maintained by the government,[91] thus drawing MYWO, a foreign donor and the Kenyan government into a single interorganizational relationship. In the mid to late 1970s, other foreign donors began to assist MYWO financially in support of projects and programs: namely OXFAM, the Ford Foundation, Pathfinder International and an individual British gallery owner in Nairobi.[92] The establishment of these interorganizational linkages between MYWO and these donors would serve as a continuation of a substantial increase in the numbers of foreign donors "interviewing" MYWO while looking for women's projects to fund during the Decade for Women (1975–85). Moreover, the establishment of these interorganizational linkages would serve to substantially increase the amount of foreign funding assistance MYWO would have at its disposal.

MYWO's successes under Kiano can also be attributed to its endorsement of and participation in *harambees*. Under Kiano's leadership, the organization had begun to generate a significant amount of money through its own local income-generating activities including fund-raisers, fashion shows, *barazas* (receptions, meetings), jumble sales and yard sales.[93] The Kianos (Mrs and Mr) were often principal donors.[94]

MYWO's move towards self-help was fostered by the spirit and ideology of *harambee*, popularized by Kenyatta upon his declaration of internal self-government for Kenya.[95] *Harambee*, which literally means "Let's pull together," became the national motto of Kenya, aiming to bring together communities to meet their own needs with their own resources. Ultimately, the goal of *harambee* was to facilitate the overall development of Kenya.[96] Thus, in support of Kenyatta's declaration of *harambee*, MYWO's focus from the national office under Kiano became local income generation. As indicative of MYWO's focus the chair of MYWO Central Province Grace Wanyeki stated, "In the 1970s we started to think of income generating projects such as pig keeping, livestock rearing, poultry keeping and many more. It was heartening the indiscriminative participation of women of all ages.[97]

Over time, *harambee* would also include the utilization of "outside" resources to meet local community needs. The Kiano administration embraced this modification and expansion of the *harambee* ideology, as it was simultaneously securing foreign donors for development projects. Unlike Tanzania, which stressed *ujamaa*, self-reliance, and rhetorically rejected external aid in the 1970s, Kenya welcomed it. Women's groups in Kenya increased in number as well at this time. This increase in number is probably a direct result of the increasing availability of aid to women's groups in the country, and the subsequent belief that women's groups had a strong chance of getting funding for development projects.

Finally, Kiano's influence in women's affairs internationally is a direct cause of MYWO successes. Kiano was key in securing foreign donors for the organization because she played such a high-profile role in women's affairs internationally. It was Kiano who brought MYWO to national repute, primarily because of her dynamism and political savvy on the international scene, and secondarily because of her personal connections to top Kenyan government officials who could promote MYWO. MYWO had accomplished much, but on those accomplishments alone it would not have secured the degree of funding assistance from foreign donors that it achieved from the late 1970s into the early 1990s.

Kiano was a mover and a shaker and she networked quite effectively in the international circles among those who could allocate large sums of money for development projects. For instance, Kiano had served as the key representative and voice of the women of Kenya and as the chief delegate of MYWO in the Women's Conference in Mexico in 1975. She was also the chief delegate in Copenhagen in 1980 where women from both beneficiary and donor countries met to discuss women's issues and to plan programs to integrate women into the development of their home countries. In very important ways, these conferences lent women from beneficiary countries the opportunity to make their case to foreign donors for assisting their particular organizations and countries. Kiano, the politician, seized these opportunities. Moreover, in 1980 at a meeting in Hamburg, Germany, Kiano was elected co-deputy president of ACWW, an organization nine million strong that has been engaged in interorganizational relations with MYWO since colonial times. Kiano is the first African woman to have been elected to this post.[98] This is only one example which illustrates the leverage Kiano had on the international women's network. She had gained outstanding recognition for herself, MYWO and Kenya from donor countries.

Kiano had many assets in her favor in attempting to secure foreign donor support for MYWO. For example, she could boast that MYWO was the largest countrywide women's organization in Kenya, with a women's group infrastructure already in place to implement programs from the national headquarters to the grassroots. She could argue that Kenyan women made overwhelming contributions to development, as women in Kenya are the agriculturists and are major contributors to the economy. She could speak to Kenya's political stability, at the time. In addition she could also boast of Kenya's phenomenal economic performance since 1963 compared to other African countries. Generally, donor countries already viewed Kenya as a capitalist success. Kiano had a host of assets from which to draw as she campaigned for donor funding for MYWO and for Kenya. This was quite satisfactory for the requirements of many donor programs, as many foreign donors look for development successes first, and then decide to put money into women's projects.[99]

The larger external donors that made long-term interorganizational funding commitments to MYWO projects and programs did not begin to fund MYWO for extended periods until the late 1970s. Of the four major long-term programs with foreign donor funding during the time of my fieldwork (1991–2), three began their implementation phases and one its formulation phase during Kiano's administration (1971–84). The Maternal Child Health/Family Planning program (MCH/FP), funded by Pathfinder International, the World Bank, USAID and the Norwegian Agency for International Development (NORAD), began its implementation in 1979; the Nutrition program, funded by Marttaliitto, began its implementation in 1980; the Special Energy Program–Jiko (SEP–Jiko), funded by the German Agency for Technical Cooperation (GTZ), began its implementation in 1983; and the Leadership Development program (LD), funded by the Konrad Adenauer Foundation, began its formulation in 1984 and its implementation in 1985. There was another program of income generation which had begun in 1984 and which was projected to be long term. Instead it ended abruptly without much explanation, other than that the Ford Foundation would not fund an NGO which had become part of a political party.[100] This income-generating program was being managed by Mrs Wilkista Onsando, the current chair of MYWO, who had resigned as the chief executive officer (CEO) of MYWO to manage the program. The program was being implemented in the districts of Muranga and Kisii, interestingly the home areas of Kiano and Onsando respectively. Until these foreign donors began providing major financial assistance to MYWO, the organization had not had programs on a large scale. They simply did not have the sufficient financial resources to implement them,[101] and thus they engaged in interorganizational exchanges to do so.

The KANU government's move towards affiliation of MYWO: key players set the stage

In an interview in 1982, Kiano was asked whether or not there was a sect in MYWO that sought a more militant approach to development which some members felt might alienate men. Kiano responded:

> Our aim as an organization is not to fight anybody and certainly not to fight men. We view harmony as the foundation for working together and understanding each other. We don't have any group whose militant approach can alienate men. We are out for cooperation and mutual coexistence is our ideal.[102]

This statement aptly described MYWO's official organizational approach to cooperation with the government for the development of Kenya. Since MYWO's disassociation with the colonial state from Asiyo's through Kiano's terms, MYWO's relationship with the government had been informal, with no specific definitions of their respective organizational duties and responsibilities in interorganizational development cooperation. Generally, it was assumed that MYWO would work with the government as a complementary interorganizational entity in the development direction the government chose. Personnel from the Kiano administration explained that this was not unusual since "most of the work of non-governmental organizations (NGOs) then was complementing the work of the government, although there was no formal coordination."[103]

The national leadership of MYWO used motivational propaganda to encourage local women to contribute to this effort, yet they simultaneously and contradictorily said that they developed their own agenda from the perceived direction of MYWO grassroots members.[104] "Cooperate with the government" was a common theme of MYWO to encourage women to participate in as many development projects as possible.[105] Although this theme was much more pronounced in the Moi government after 1978 than it was in the Kenyatta government, women's call for involvement in development after independence originated with Kenyatta (1963–78).

Though Kenyatta used rhetoric favorable towards women's involvement in Kenya's development, there was little encouragement by his government, in the way of investing resources in women, individuals or groups, to facilitate women's involvement in development in any meaningful way. For example, *Women of Kenya, A Review and Evaluation of Progress* reveals that in 1975/76, during Kenyatta's presidency, only 0.54 percent of the total Ministry of Culture and Social Services (MCSS) recurrent budget was allocated for direct support of women's activities—a total of Kenya £8,400 (US $20,265) for women's groups with a membership of over 156,892 women.[106]

Generally, during Kenyatta's administration, women as a group were hardly mentioned in development policy or in other major policy statements.[107] The government did not feel it necessary to mention women or acknowledge their contributions to the national economy,[108] as their role in development activities was not regarded as official business of the state. Generally, women's contributions tended to be seen as unofficial and informal, and hence less important than men's contributions. Although women's contributions were critical to the economy, they were regarded as peripheral. Women and their labor were exploitable. Kenyatta's personal dealings with an MYWO women's group is a case in point.

The Thika Township Women group of 150 women organized in 1971 and set a goal to buy some land to grow coffee. There were problems because the group did not have the means to raise all the money necessary for the purchase. After the women searched for alternatives to no avail, they were hired by Kenyatta to cultivate his garden. In exchange, he paid the women Ksh 2 per day (less than 5 US cents) for their labor. After some time, the group was forced to increase the membership of the organization by 700 members, from 150 to 850 women, in order to get enough money to purchase the land.[109] The government, however, maintained that it was "not aware of overt discrimination against women in the country."[110] It argued that Kenyan women were not discriminated against and therefore did not need to struggle for rights they were already enjoying.[111] The Kenyatta government was able to personally and publicly exploit both the fruits of women's labor and their contributions to local and national development through their participation in *harambees*. MYWO not only allowed, but also encouraged this exploitation.

Women in the rural areas were in fact the backbone of the *harambee* movement, and as such their overall contributions to national development far outweighed government contributions and foreign donor contributions. For instance, John Orora and Hans Speigel, in a study of *harambee* projects in Kenya between 1967 and 1973, demonstrate that the lion's share of the monies for *harambee* projects came from local people.[112] In their study they found that about 90 percent of contributions for *harambee* projects came from *wananchi* (ordinary citizens). Comparatively, they found that of a total of Ksh 382,000,000, the central government contributed only 20,000, local authorities 3,000,000 and other donors, including foreign, only 12,000,000.[113] The local people—mostly women, though not identified as such by Orora and Speigel—had raised and contributed almost all of the monies for the projects. Contributions by government and foreign donors for development cooperation were negligible. In addition, in 1976 Kenyatta acknowledged the contributions of *harambee* projects, announcing that local people had contributed "40 percent of capital development in the rural areas" as well as making voluntary cash contributions of Ksh 1,000,000 in 1976 alone (approximately US $13 million).[114] Although Kenyatta did not speak of women specifically, it is a well-established fact that women are the bulk of the rural population and were in fact the main contributors to this effort. Kenyatta used MYWO, specifically rural women members as well as non-members, with the leadership's complicity, to meet the development demands that Kenyans made on the state.

Kenya's 1989–93 National Development Plan, written during the Moi administration, demonstrates that the government's contributions to *harambee* development projects have remained consistently low over time. For example, although the total value of *harambee* contributions had increased by almost 400 percent from Kenyan £9.79 million in 1979 to Kenyan £37.29 million in 1985, the Kenyan government had contributed on average only 9 percent of the project costs over the seven-year period.[115] It is important to note the mention of women in this Development Plan of 1989–93. The plan was written after the Decade for Women when the mention of WID (Women in Development) was fashionable and expected, and when its neglect carried the stench of a major international sexist social faux pas. With specific regard to women and *harambees*, the plan stated:

> Women's group participatory initiatives have also contributed tremendously to economic and social progress in the country. Besides their contributions in cash to *Harambee* projects, they have also made major contributions in terms of materials and labor particularly because they constitute the majority of the rural population. Through various women's organizations in the country including *KANU–Maendeleo Ya Wanawake* [emphasis added], National Council of Women of Kenya [NCWK], the Women's Bureau [WB], the Young Women's Christian Association [YWCA] and the Girl Guide Movement, there has been growing and effective participation of women in developmental activities across the country. Such organizations will continue to receive encouragement and support during this Plan period.[116]

On the surface, it seems that the mention of women in the 1989–93

National Development Plan is progressive. But this, in and of itself, does not indicate that women have been fairly acknowledged or integrated into national development planning in any significant way. On the contrary, during the Moi administration, at which time this plan was written, women remained very marginally integrated in official strategies for development. Moreover, when women were marginally integrated, they were very often supervised and censored by men, as in the case of MYWO. Beyond broad statements of recognition and praise for compliant women and the consequent "good" intentions of government to support women, there is very little substantive policy or genuine commitment to assist women. With this in consideration, the statement in the National Development Plan merely indicates that the Moi government recognizes the role that MYWO (and other women's organizations) play in the mobilization of women's labor for *harambees*, and publicly calls upon them to continue to host *harambees* to raise more money.[117]

In his book *Kenya African Nationalism: Nyayo Philosophy and Principles* Moi deceptively gave the impression that women were highly regarded in the development policies of Kenya. He posited, "The Government of and people of Kenya will give the progress of women the highest priority. Their progress is our progress; their stagnation is our backwardness."[118]

Further on, however, he qualified this assertion. In a chapter devoted to women, notably separate and apart from the chapter on strategies for national development, Moi indicated that, "Further progress for the women will be achieved only by continued cooperation with the Government, in faith, loyalty and co-action with the leadership from the grassroots to the top."[119]

Moi further admonished women not to measure their progress in terms of rebellion and confrontation with men, but instead on the basis of cooperation with them.[120] That is, women should pledge their loyalties to the government—to men—and seek their permission and consent for their actions to ensure their continued individual and organizational progress. With specific regard to MYWO, Moi subtly announced what was to come—the affiliation of the women's national organization to the KANU ruling party and the KANU government. Moi justified this impending merger by stipulating that MYWO's affiliation with KANU would "let it (MYWO) evolve into a new stage of dynamic maturity."[121]

Moi, who had been vice president to Kenyatta, became president of Kenya in 1978 upon Kenyatta's death. Although Moi was versed in the post-1975 Women in Development (WID) rhetoric and used it often, the budgetary allocations during his administration for women's programs demonstrate the impotence of his words and his general disregard for women's activities. Between 1978, the year that Moi assumed the presidency, and 1982, the government allocated the equivalent of 0.1 percent of the total government expenditure to women's programs. Moreover, since that time government grants to women's groups have consistently dropped, from Ksh 3.3 million in 1986 to Ksh 2.6 million in 1987 to Ksh 1.7 in 1989.[122] Moi's strategy has been to wait for women's groups to capture money for the state. He managed this through the affiliation of MYWO in 1987.

Both Kenyatta and Moi had attempted to officially affiliate MYWO to KANU—the party and the state—when Kiano was chair of MYWO. Both presidents had waged private, intense battles against Kiano to make MYWO an official part of the state apparatus. Kiano's cohorts revealed that one of the main reasons she remained chair of MYWO for 13 years was to prevent this affiliation. She had successfully resisted for 12 of those years.[123] In April 1984, however, in large part because KANU could no longer be prevented from affiliating MYWO, Kiano resigned as chair of MYWO. She resigned three years before her active term was over, to everyone's surprise, on the eve of the UN Decade for Women meeting in Nairobi in 1985.[124]

Asiyo, the first African woman chair of MYWO, may have been aware of this pressure, as she forewarned in a 1982 interview:

> There have been a lot of handicaps and bottlenecks but Maendeleo is the only organization that has succeeded against a lot of difficulties both before and after independence. It is my sincere hope that nobody will attempt to meddle with it. It would be a shame if anybody discouraged Maendeleo from aiming even higher. We have come so far and Maendeleo will live forever.[125]

KANU had in fact "meddled" with MYWO. KANU's affiliation of MYWO was most clearly an attempt to increase state control over non-governmental organizations (NGOs) in Kenya, specifically the largest women's NGO. Many wondered why KANU chose to affiliate MYWO over other organizations such as Central Organization of Trade Unions (COTU), Law Society of Kenya (LSK), National Council of Churches of Kenya (NCCK) and NCWK. There are a multiplicity of reasons for this.

1 MYWO was perceived as an organization which could be particularly useful to the state. MYWO reportedly had a membership of over 1.5 million women who could be co-opted supporters for KANU. Its network, though weaker than during colonial times, stretched between the national headquarters in Nairobi and the rural villages, thus offering potentially useful support to KANU candidates at various levels. Moreover, it was commonly known that the women's votes determine who goes to Parliament,[126] and so it was politically important to KANU to capture these votes.
2 MYWO could further assist with the implementation of the decentralization initiative, which Moi announced in October 1982, since many of the women were already involved in rural development. The women had

already demonstrated their contributory potential to national development and their commitment to nation building. They were a pool of cheap labor who could implement low-cost development projects with minimal government contributions.[127] KANU's rationale in this matter was clear—by completing rural development projects which serviced local areas, the women fulfilled a tremendous amount of the state responsibility, and thus cushioned and mitigated citizens' demands and frustrations towards the state.

3 Additionally and extremely important to KANU, MYWO could attract various foreign donors to Kenya to fund projects with particular emphasis on women. MYWO had been particularly successful in securing millions of dollars under Kiano. For example, for at least two projects for which funding had been secured under Kiano, the chief executive officer (CEO) of MYWO, Mrs Jane Kirui, announced in 1990 that the organization would spend Ksh 50 million in 13 districts.[128] Thus, by being affiliated to MYWO, the party-government and state would capture MYWO's purse strings and have power to control MYWO's finances and spending decisions. For a country whose external aid debt had reached Ksh 154 billion (US $5.5 billion) at the end of 1990, this was important.[129]
4 MYWO was considered particularly vulnerable by the male-dominated KANU because it is a women's organization. By virtue of its femaleness, it was considered to be less forceful, less resistant to being overtaken and easier to dominate. As one interviewee stated, "MYWO was KANU's cheapest hunt."[130] KANU had in fact tried to affiliate COTU and LSK, but to no avail. Both organizations, which were traditionally male-dominated, had resisted the affiliation. KANU would thereafter try to dominate those organizations which it believed to be weaker.
5 MYWO had also proved itself to be the most accommodating and least threatening to the government of all women's national NGOs, taking conservative to moderate non-confrontational stands on women's home-making issues. Because of the MYWO's posture, the government believed that MYWO could be used, as it was used, to silence other non-compliant "unruly" women's organizations, particularly NCWK. The government had a personal vendetta against NCWK's leader, Professor Wangari Maathai, who had consistently challenged the government on its blatant violations of human and environmental rights; and who had been victorious. MYWO would become the mouthpiece of the government to try to control Maathai and other women in Kenya.

It was Mrs Theresa Shitakha, who became Kiano's successor as chair of MYWO in 1984 and who had been Kiano's vice chair and a member of the MYWO national executive committee, who set the stage for MYWO to eventually assume the role of appendage to KANU. Shitakha's brief chairship (1984–85) was marked by allegations of inefficiency, financial mismanagement, corruption and nepotism which provided the near ideal circumstances for KANU's more direct intervention into MYWO's affairs. Problems and confusion surrounding Shitakha's chairship and the eventual decline of MYWO began before the Nairobi Decade of Women Conference (1985). Although the conference went on smoothly, problems loomed large.

In October of 1985, shortly after the conference, and one year and six months into Shitakha's chairship, the *Standard*, a leading Kenya newspaper, made public the allegations of inefficiency, financial mismanagement, corruption and nepotism against the chair and her national committee. The newspaper further asserted that MYWO was in a "financial quagmire", owing the Commissioner of Income Tax nearly Ksh 6 million.[131] These allegations would begin a process of unraveling MYWO and altering its interorganizational relations with the government and foreign donors.

Kenneth Matiba, who was then the Minister of Culture and Social Services, and Eliakim Masale, the permanent secretary, along with three other senior officials, promised to create a government probe committee to "unearth the truth" with regard to these allegations against Shitakha.[132] This committee was to investigate the following:

1 a tax debt of Ksh 5.9 million;
2 failure to account for expenses incurred for the project "Information for Rural Health/Family Planning (IRH/FP) in the amount of Ksh 145,000";
3 payments for goods not tendered;
4 tribalism and sackings.[133]

By mid-December 1985, Shitakha and her executive officers were suspended.[134] Before the end of 1985, two senior officials were also ordered to stop using the MYWO vehicles and turn the keys over to the accountants. Stipulations were made for the MYWO vehicles to be used only for official functions and finances to be spent only for official expenses.[135]

The MCSS probe found evidence to support all of the allegations of inefficiency, financial mismanagement, corruption and nepotism. It concluded that MYWO's financial undertakings were in need of complete reorganization.[136] KANU, the party, would step in to assume the responsibility.

Shitakha was immediately dismissed as chair of MYWO and replaced by MYWO Coast Province chair, Mrs Mary Mwamodo. Mwamodo had been a member of MYWO since 1956. Shitakha's chief executive officer, Mrs Gladys Mulindi, was also replaced by Mrs Dorcas Kamau, a long-standing member of MYWO and the LD program officer 1985–91. It was expected that Mwamodo would serve until the new MYWO national elections were held. Mwamodo announced the polling would be April 1986, with no specific date.[137]

Shitakha challenged the findings of the MCSS probe. She insisted that she was still the undisputed leader of MYWO despite the findings of the committee. She filed an injunction with the high court against MCSS, which was thrown out.[138] Shitakha claimed that she was a scapegoat, a fall guy. She argued that when she took over as chair of MYWO the organization already had "money problems and this became an obstacle to many projects."[139] She implied that these were problems of which Kiano was acutely aware and which she passed on. Kiano and her treasurer who succeeded Dar, Mrs Florence Gichuhi, had resigned just prior to the surfacing of these problems. Gichuhi indicated that she had resigned in 1985, not long after Kiano, because "when Shitakha took over, she misused money. When Mrs Kiano left, money began to be used for many unauthorized things. There were many bills."[140] Gichuhi insisted that she did not want to be involved in a financial mess.

Chaos gripped MYWO as it split into factions, many of which looked to the government for direction. The KANU government immediately seized the opportunity to direct MYWO out of its furor. It kicked out Shitakha's entire executive committee, postponed the MYWO polls from April 1986 until June 1987, and appointed a caretaker committee to be headed by civil servant Mrs Francisca Otete to run MYWO's affairs, commencing May 1986.[141]

A review of MYWO and KANU correspondence between 9 January 1987 and 16 April 1987 reveal that the affiliation had taken place before that time. A working committee of KANU and (K)MYWO redrafted the MYWO constitution into the KMYWO constitution, effectively changing MYWO's status from a non-governmental organization (NGO) to a "women's development body of KANU,"[142] hence constitutionally altering the separateness of the women's organization in the transnational interorganizational network under investigation. Despite the fact that this constitutional change had been made and MYWO's status had been radically changed, MYWO continued to claim, especially to foreign donors, that it was still an NGO. The reason is that it wanted to continue to be eligible for NGO funding from foreign donors. This claim that it remained an NGO was a desperate attempt to maintain the type of interorganizational linkages it had had under Kiano, as it had linked with its partners as an independent organization. It was a desperate attempt to continue to receive funding from foreign donors.

The KANU/MYWO working group determined that MYWO owed a total of nearly Ksh 12,000,000 to a long list of creditors.[143] Hence, they made the case that because MYWO could not pay its bills alone, it needed KANU's assistance. KANU wanted affiliation. MYWO did not resist. Had MYWO forgotten the power of women, of rural women, who had in 1976 alone, through *harambees*, raised US $13 million (Ksh 100 million)? That certainly would have been enough to pay off MYWO's debts. Perhaps the KANU/MYWO working group did not realize the financial potential of women in Kenya precisely because the former were *not* rural women.

By affiliating MYWO to the ruling party, KANU had assumed the role of "savior" of MYWO, bailing it out of its financial quagmire and appointing its leadership to mitigate its political squabbles. In exchange for KANU's assistance to MYWO, KANU had gained control over MYWO in significant ways which would change its organizational character and overall interorganizational operations. These changes would have far-reaching effects for the dynamics of interorganizational relations and development in Kenya. No longer was the transnational network trilateral between MYWO, foreign donors and the Kenyan government; it was instead bilateral—the state and its substructures, including MYWO on one side, and foreign donors on the other.

Most importantly, MYWO had compromised its autonomy by becoming KMYWO. Organizational constitutional changes had given KANU the power of voting, overseeing, censoring, condoning, defining or rejecting (K)MYWO activities—locally, nationally and internationally. KANU had also assumed the power to approve or disapprove of (K)MYWO's internal and external affairs.[144] (K)MYWO had exchanged its fate as an NGO for affiliation to the ruling party and the government. It was now inextricably tied to the party and the state. Its interorganizational linkages necessitated reconfiguration at this juncture.

A critical look at the relationship between KANU and KMYWO suggests that the events which transpired prior to the affiliation, specifically those between Shitakha and Matiba which to some extent legitimized the affiliation, were part of a plan by members of KANU and MYWO leaders to 1) solidify the state and the patriarchy; 2) strengthen the ruling party; and 3) protect African bourgeois and petty bourgeois interests. Kiano is a case in point. Although Kiano had declared that "Maendeleo is in my blood and I will always be involved in development activities,"[145] and although she remained patron of (K)MYWO, she was unusually quiet throughout the tumultuous events which led to the actual affiliation. This is problematic, particularly since she vehemently opposed the affiliation for 12 years.

Foreign donors were not very vocal as they observed the affiliation take place. Many of them watched anxiously as the dynamics of interorganizational relationships changed. Some, though not all, foreign donor funding projects did not buy the line that KMYWO was still an NGO. The Ford Foundation and Marttaliitto are examples of foreign donors that did not accept this claim. After the affiliation, the Ford Foundation withdrew their funding for income-generating projects. They argued that they could not support a political party, which they would in effect do if they continued to fund MYWO projects.[146] In addition, Marttaliitto, upon the instructions of the Finnish government, could not renew their Nutrition program with KMYWO, as KMYWO was now part of a political party.[147] Moreover, foreign donors did not feel particularly comfortable in knowing that KANU would be the "supervisor" of KMYWO's development funding and spending, given KANU's shady history and lack of credibility. Thus, the nature of the interorganizational relationships between KMYWO and its foreign donors had begun to change. The perception of many foreign donors influenced their involvement with KMYWO at the national headquarters, and ultimately with women at the grassroots. KMYWO at the national headquarters may have made some short-term gains as KANU "saved" it financially, but women's losses in the long term would far outweigh the short-term gains.

By KMYWO's affiliation to KANU, the women's organization had come full circle. After 25 years of functioning as an independent women's organization since the African women's leadership coup and their disassociation from the colonial state, KMYWO reverted to its initial status as an appendage of the state. The difference was that this time KMYWO would be directed and controlled by the Kenyan "independent" state for its own ends. The KANU/KMYWO working group even considered returning to the Kenya Institute of Administration (the old Jeanes School), where Shepherd and the colonial government had used African women in attempts to crush Mau Mau, as the venue for their workshops.[148] Because of KMYWO's financial problems and its perceived inability and lack of resources to solve those problems, KMYWO was particularly vulnerable to and dependent on the KANU party and the state—at least, its leadership gave this impression. In its interorganizational relations, KMYWO had been coopted by a larger and more powerful organization in its network, which was perceived to have more resources at its fingertips, and thus would be able to keep MYWO afloat. Of course, this was at a significant cost to MYWO in that its organizational autonomy had been severely compromised.

MYWO's status as a benefactor of development and asset to the state had changed as well. That is, KMYWO, with the possible exclusion of the leadership—themselves extensions of the state—was tricked into believing that the organization was a beneficiary of the state to be aided by KANU. As such, KMYWO would allow its labor, finances and political patronage to be exploited by KANU in interorganizational exchange for KANU's paternal assistance and "protection." There is no doubt that KMYWO felt beholden to KANU. Ironically, KANU was benefiting more from the affiliation than (K)MYWO grassroots women. KANU would now have control of the foreign donor assistance MYWO secured.

Kenyan women had been used by the colonial state through MYWO. They revolted, however, and won the war. Now they were entrapped by the post-colonial state in a battle that was more protracted.

PART II: REACTIONS TO AFFILIATION AND THE KMYWO 1989 ELECTION

Initial reactions of KMYWO and KANU leaders to the affiliation were overwhelmingly positive. For example, KMYWO leaders and KANU women leaders gave the public the impression that MYWO's problems had been solved by "chivalrous" acts of KANU, and that women's status and power had been enhanced by the merger. Otete, the chair of the KMYWO caretaker committee said, "The merger has helped in many ways, one of which is the exemption from income tax payments for the Maendeleo house."[149] Dr Julia Ojiambo, then the KANU Director of Women and Youth Affairs and former Assistant Minister and Member of Parliament, said that the affiliation would secure representation for women on decision-making committees in all development sectors in the country. She claimed that KMYWO was under her jurisdiction and as such the merger would ensure that women's issues would be closer to the government's ear.[150] Ojiambo agreed with the patriarchal KANU government on the dependence of women. She justified the affiliation by saying, "They [women] have more chances of developing, as they are working in close collaboration with and partnership with KANU."[151] In an editorial in *Viva* magazine, which since the affiliation included men on the board, the affiliation was referred to as a positive step towards the "regeneration of women's activities," from which KMYWO and the country would benefit. *Viva* declared the merger between KANU and MYWO to be "a marriage full of promise."[152] There was overall optimism among the leadership, particularly since Mr David Amayo, the then national chairman of KANU, pledged that KANU's relationship with KMYWO would be "one of non-interference."[153]

Nothing was further from the truth. KANU's interference with MYWO became so heavy-handed that KMYWO leadership could not be distinguished from KANU leadership. Men assumed the running of the organization by making up the rules for governing the women's organization under the auspices of KANU—the party and the government—and through women puppets. As with KANU, whose entire national leadership is male, men were in control of KMYWO. This was most evident in the national elections of KMYWO, held between 30 October and 4 November 1989. Moreover, KANU was the cause of the election's eight postponements. KMYWO elections were initially scheduled for April 1986 after KANU's intervention into MYWO affairs, but amid continual organizational problems with MYWO's affiliation to KANU, elections were not held until nearly four years later. It was not uncommon for newspaper headings to read: "Maendeleo elections postponed"; "Maendeleo polls postponed again"; "Maendeleo's elections postponed once again"; "Maendeleo date changed"; "KANU Maendeleo polls today."[154]

The reasons cited for the postponement of KMYWO elections in August

1989 include the national census (the first taken since 1979) and the Mombasa Agricultural Show, which was scheduled for August. KANU argued that these two events would have reduced the registration time for women voters and hence would have affected the outcome of the elections.[155] The Nairobi International Show, scheduled for 27 September, and Moi and Kenyatta Day celebrations scheduled for 10 and 20 October respectively, were again reasons for postponing the elections in September 1989.[156] KANU national chair Peter Oloo Aringo announced that finally KMYWO's poll would be held in late October after a "breather" from these events.[157] KANU very strategically utilized the time during the postponements to further plan the women's elections and to recruit voters for KMYWO and KANU primarily. It was argued by some that this was the real reason for the postponements.[158] KANU announced that women had to show their KMYWO cards in order to vote, and their dues had to be paid up to date.[159] Pending the election, Ojiambo, the director of KANU Women's Wing (of which KMYWO was *not* part), and Aringo announced that 1.12 million registration cards had been sold to women throughout the country.[160] A little over one million cards sold to a total population of ten million women, as estimated by the Central Bureau of Statistics, is hardly an astounding number.

During this time, KANU men began charging each other with interference in the women's elections in attempts to end each other's—the men's—political careers. Rivals within the party accused each other of wooing particular women with the intent of forging political allies. Women were being used as pawns in men's political games. The editor of the *Daily Nation* observed:

> Some men were claiming there were fake party stamps from within and outside Maendeleo being distributed (to women) on a selective basis. Men were accusing other men of dishing out huge sums of money in various parts of the country in a bid to influence the women's elections. . . . Men were accusing the male branch chairmen of selling KANU stamps to women on a selective basis. The powerful men of the party branches were issuing their own versions of the procedure for the women's election issued by the headquarters.[161]

Other issues contrived by KANU men became part of the melee as well. For example, Mulu Mutisya, Machakos district KANU chairman and MP, warned that there were "some women" native to Machakos but who now lived elsewhere who were planning to contest the elections. He advised his constituency that "Such women should be rejected during the elections since they are not acquainted with the problems the district residents have."[162] Councilor Tony Ndilinge, Machakos District youth leader, also alleged that "some people were planning to interfere by bringing candidates of their choice from outside the district."[163] From another area, the Narok District commissioner and KANU member Mr John Sala stressed to Narok chiefs and KANU officials the importance of ensuring that all members of KMYWO were registered so as to facilitate their full participation in the elections.[164] Even President Moi became directly involved in the women's elections. He warned women not to be "cheated" and "bought," and he alerted men to "keep off the polls."[165] However, he intervened. Four days before the elections, contrary to other election requirements previously made, Moi announced that the only requirement for women to vote in the KMYWO elections was that they should have their national identity cards. This meant that KANU party election rules were turned on their head—a woman did not have to register as a KANU member, neither did she have to be a member of KMYWO in order to vote. Moi argued that he ordered this action to avoid confusion brought on by fake membership cards and the intentional maldistribution of membership cards.[166] By making these pronouncements, Moi had succeeded in undermining those KANU members and politicians who had themselves schemed to undermine the women's election. He had also cleverly increased his own popularity among women at the grassroots, who may not have met the requirements beyond the national identity cards. Moi's order, however, did not stop the confusion.

Men were determined to participate in the women's elections at all costs. Perhaps their most bizarre plot was revealed by the KANU district officer for Kiamaa Division in Kiambu District, Mr Haroun Ichima. Ichima revealed that there was

> a well organized plot by some men to not only influence the outcome (of the elections), but to actually participate in a more direct way. The men planned to pose as women by wearing dresses and stuffing oranges down their shirtfronts in order to pass muster as bona-fide women voters.[167]

Although no men were found with oranges in their shirtfronts on election days, men were front and center at the elections. Many women boycotted elections because of men's meddling. For example, at the sub-national levels, in Nyeri District a group of women refused to vote in protest at men's interference in the polls. In Dandora Ward, Nairobi, a group of women forced a truck carrying women voters off the road. They claimed that these women were being transported to the polls by a senior politician from the area who had bought their votes. Mrs Lydiah Wanjiru, leading the women protesters, said, "We cannot allow these elections to be tampered with. We want our elections to be for the womenfolk alone." Another group of women at Ngong claimed that KANU officials had given *chai* (bribes) to the women to vote for KANU's candidates. They also named politicians from other areas who had poured money into Quarry Ward, Ongata Rongai Ward, Maasai Ward, Lenana Ward, Scheme Six Ward, Kisumu Ndogo Ward and Ngai Murunya Ward.[168] A group of women in East Wanga Location, Kakamega

District, also protested at the interference of KANU men in the elections. As a result, the chief of the location, Mr J.A. Kodia, ordered the police to arrest the local KANU chair for interference.[169]

Perhaps the most flagrant of all KANU officials who tampered with the elections was Mombasa's Mr Shariff Nassir, Assistant Minister for Information and Broadcasting. Nassir was accused by the former Mombasa District MYWO leader, Mrs Zubeda Sumba, of openly campaigning for "a particular woman of his choice." This woman was his sister, Aisha Shariff. Sumba and others also charged that Nassir openly dished out money in the campaign.[170] Nassir had attempted to disguise his moves by calling a press conference two weeks before the elections to publicly state his position against men's interference in KMYWO's elections. He stated, "There should be no interference by men at any quarters." Again, prior to the elections, he held another press conference warning his political rivals not to show up at the polls. It was Nassir, however, and not his rivals, who was spotted at two polling places in Mombasa—Majengo and Shimanzi.[171] It was Nassir who reportedly mobilized KANU youth wingers to bar women who opposed his sister from the polls.[172] Between 8,000 and 10,000 women demonstrated along Mombasa streets after boycotting the elections because of Nassir's blatant interference. Singing songs of protest, they appealed to Moi to nullify the elections and called for a repeat of the polling.[173]

Men had demonstrated quite obviously that they had a vested interest in the polls. They had planned the elections, spoke on behalf of women, argued with each other, attempted to vote with women, bought votes, lied and schemed. They had demonstrated that they would go to far-reaching lengths to be in control of women's activities and to maintain control of the former relatively independent women's organization.

At the national level elections, men also attempted to influence and control the elections, but more shrewdly and surreptitiously than at the sub-national levels. The *Weekly Review* reported, "Male politicians from all over the country, present as observers, were very much in evidence at the Nyayo national stadium, apparently giving instructions to the female delegates as vigorous lobbying for the various posts at stake continued."[174]

Men as well as women were anxious to see the face of KMYWO for the next three years, as well as to speculate on the nature of the partnership it would have with KANU. Although the men were restrained in overtly influencing the women, kinks in the women's negotiations demonstrated the extent of the role KANU senior politicians had played as "advisors" to the women. For example, Kiambu District KMYWO chair Mrs Rose Waruhiu led a protest against the Central Province women's delegation for not including a Kiambu representative. It was KANU men officials who intervened to break the deadlock.[175]

Another event which adds suspicion to the extent of involvement KANU politicians had in KMYWO's elections was a slip of the tongue by the

Nairobi provincial commissioner, Mr Fred Waiganjo, who officiated at the elections. Without the official nomination forms for candidates in his hand—Ojiambo claimed that she forgot them on election day—Waiganjo began to announce the names of the nominees. The point is, he should not have been aware of the names of the nominated, since they had not yet been revealed. However, Waiganjo began to read the name of the nominator, Mrs Ester Wandega from Western Province, and her nominated sole contender for the chair of KMYWO, Mrs Wilkista Onsando, from an unofficial list of names in his hand *before* realizing that Wandega had not yet introduced herself to him or revealed the name of the nominee.[176]

That men had taken an interest in women's elections is an understatement. They were the most prominent figures in the event, causing much more of a "squabble" and "ruckus" than the one they took credit for resolving within MYWO. The women, on the other hand, did not appear very excited about the organization or the elections. As the daily and weekly news indicated, until the eve of the election no single woman had declared her interest in the top post of the organization. Moreover, by election time only four women had declared any interest in running for office. Furthermore, the offices in which they expressed interest were the lower-level, sub-national offices.[177] Undoubtedly, MYWO's troubles of the immediate past had led to women's lack of interest in running for office. MYWO's merger with KANU had confused and dismayed many women to the extent that they refused to participate in elections which turned out to look like a sham. The consistent postponements of election dates and the constant changing of election rules by KANU, especially the requirement of queue voting, further caused women to lose interest. Even "prominent female politicians and activists who had indicated a wish to vie for the leadership post in the first (scheduled) elections were obviously absent from the polls."[178] Clearly, KANU's persistent entanglement in MYWO's affairs and its characteristic hawking were major disincentives to women's participation in the KMYWO elections. As expressed by one woman who seemed to sum up many women's views about the elections and the events surrounding it, "This is a women's show and they should stage it in their own style."[179]

The elections resulted in the following women taking the national offices:

Chair	Wilkista Onsando, Nyanza Province, Kisii District
Secretary	Mareso Agina, Rift Valley Province
Treasurer	Florence Maingi, Eastern Province
Vice Chair	Nelia Githeka, Central Province
Assistant Secretary	Phoebe Alouch, Nyanza Province
Assistant Treasurer	Joan Mjomba, Coast Province
Executive Officer	Jane Kirui, Rift Valley Province

On election day the KMYWO national executive committee appointed the executive officer. All of these women elected were KANU and MYWO members; some of them life members of both organizations.[180] Mjomba had the longest membership in KANU, and as elected national assistant treasurer of KMYWO she had no qualifications to serve in that post. Onsando's husband had previously served as MP for their home area in Kisii. During Kiano's chairship, Onsando also had connections with KANU as MYWO's CEO under Kiano before resigning to take the post of Program Officer for Income-Generating Activities supported by the Ford Foundation until its withdrawal of funds upon MYWO's affiliation with KANU. Onsando had worked first hand with the development cooperative partners—MYWO at national headquarters and grassroots, the KANU government and foreign donors.

Although they may not have held posts in KANU, as it is an unwritten requirement that to be in the party hierarchy one must be male, all of the women who took KMYWO national offices were considered important women in KANU politics. That is, they assumed the highest unofficial complementary roles women could—"assisting" the party. Finally, through them, women's importance could be "legitimized" by their attachment to KANU, even though they represented a subordinate and unequal women's body in the party.[181]

Outside of the elected leadership, many women were enraged at KANU men's interference in the polls. In protest, they did not vote and they further appealed to their elected male representatives and to Moi for some type of redress. These appeals were in vain since their representatives were KANU members, most of whom would not rock the boat. Moreover, the opinions of many of the rank and file members of KANU mattered little anyway. One group of women referred to the entire MYWO elections process as a "defeat (of) the process of democracy—linking the elections sham to the widespread corruption in Kenya and to foreign donors call for democratization."[182]

Not surprisingly, the calls to nullify the elections and host new ones fell on deaf ears. One can but agree with Nzomo in her assessment of the situation. She states, "Despite ordinary women's cries of 'rigging' and 'male interference' the powerlessness of the majority of women in (MYWO) was again reaffirmed."[183] Ordinary women's—that is, grassroots women's—political proximity to KANU was not close enough to matter on the Kenya political front. The leadership of KMYWO, whose voices might have carried more weight, had been coopted and silenced such that their political fate was entrusted to KANU, the patriarchy these women would not dare to defy.

Following the KMYWO's elections Moi made an announcement which would further seal the fate of the organization. He stated that all foreign donations to women's groups in Kenya would, from that point, be channeled through KMYWO.[184] That is, KMYWO would receive all of the monies for women's groups in Kenya, whereas before the affiliation and the elections, foreign donations went directly to the independent women's NGOs.

In effect, Moi was taking away the independent status of all women's organizations and linking them to the party and the government through KMYWO. MYWO had been reconfigured into an umbrella women's organization, and was being used as a tool to usurp the power and independence of all women's groups in Kenya. Moreover, Moi was placing KANU in control of all women's groups' purse strings. All of their foreign funding would now be overseen by KANU, and would be directed, controlled and utilized by KANU, for KANU's best interest. Most importantly, Moi was placing the KANU government at the top of the interorganizational development partnership hierarchy, in charge of KMYWO and foreign donors, whose resources it had stealthily garnered and would now exploit.

Post-election changes and disintegration: KMYWO's relations with the government and foreign donors

KMYWO's first meeting after the elections suggested, very loudly, that KMYWO would not be exerting much autonomy as an "independent entity" of KANU. Instead, the meeting suggested that KMYWO would be dominated by KANU's continuing paternalistic control.

KMYWO's first meeting was chaired not by its new chair, Onsando, but by KANU's Director of Internal Audit, Mr A. M. Aburi, who is said to have analyzed MYWO's performance in 1986–7 under the caretaker committee. Aburi reported that 1) MYWO did not have an "enlightened leadership, an adequate management system or an accounting system"; 2) MYWO never used funds for the purpose for which they were intended, and some funds were fraudulently misappropriated; and 3) more than 60 percent of the organization's annual income of Ksh 33 million was used to pay salaries.[185]

Otete, the chair of the caretaker committee, vehemently challenged Aburi on his facts, figures and ability to conduct a thorough and accurate study. She was, however, silenced by Kiano, who by her actions had symbolically given her support to Aburi and KANU and had discredited Otete.[186]

Aburi made the following recommendations for KMYWO which would cement KANU's partnership with KMYWO, clearly establishing KANU as the senior partner. Aburi recommended that although KMYWO was an "independent" organization it should be overseen by KANU's Directorate of Youth and Women's Affairs. This directorate would be responsible for coordinating KMYWO activities, and controlling its foreign funds. It was a tremendous task KANU was willing to take on, as all of the monies in foreign financial assistance given to any and all women's groups in Kenya would be overseen by this directorate.

By these actions, KANU was entrenching its control over KMYWO, as the leader of the interorganizational development partnership. KANU was establishing rank—placing itself at the apex to direct its subordinates, KMYWO and foreign donors. In order to do this, KANU was replacing the old guard of MYWO with its own people, especially in matters concerning finance and development planning. KANU was determined to control KMYWO finances and activities agenda. Through its interactions with KMYWO, KANU was also indirectly speaking to foreign donors of women's development projects. The messages were mixed, and may sometimes have seemed contradictory, because KANU was speaking out of both sides of its mouth. On one hand, KANU was insisting that MYWO was still an NGO, maintaining its autonomous nature despite its merger with the party, and as such KMYWO was still eligible for foreign donor assistance for women's development projects. On the other hand, KANU had affiliated KMYWO to the ruling party, making it an organ of the party and compromising its autonomy as an NGO. KANU wanted financial and political control of the women's groups and all of the foreign monies they were able to solicit. At a very basic level, what KANU really wanted was to have Kenyan women secure foreign donor funds on the international market under the guise of support for their development work, while the KANU party—men—sat and waited for the women to turn the money over to them. KANU's plan was to capture the money after the women's work secured it. In exploiting women and their labor, KANU had made a mockery of Kenyan women and of their remarkable contributions to Kenya's development. Moreover, KANU had assumed the role of a madam over Kenyan women.

Some of MYWO's past leadership, most notably Jane Kiano, seemed to be in active compliance with KANU in these efforts, urging KMYWO's elected officials to work closely with government officials.[187] The recently elected leadership seemed to be in complicity as well. That is, Onsando and her national executive appeared to be KANU puppets in silent acquiescence. Perhaps Aburi's performance at KMYWO's first meeting was staged to show Onsando what would happen to an unruly and rebellious chair of the KMYWO. Clearly, the affiliation between KANU and KMYWO was not a marriage of equals.

KMYWO's relationship with KANU after its first meeting reflected the hierarchy which had been established. KMYWO, while still claiming a non-political autonomous nature, became the politicized echo of the government, particularly to foreign donors, multiparty proponents—domestic and foreign—and Kenyan political activists. As an opening to her chairship, Onsando announced to women's groups who were reluctantly merging under the umbrella of KMYWO that "No differences exist between the party, KANU and women's organizations."[188] KMYWO became the defenders of the KANU government, Moi and Kenya's one-party state.

Onsando was often quoted in the newspapers between 1990 and 1991, speaking to the "unreliability of foreign donors" for women's development projects.[189] She defended Moi's dictatorial practices against multipartyism, as well as the KANU government's widespread corruption. She took on the international community, which was bringing pressure on Moi and KANU, by calling them "foreign meddlers out to destabilize the country [Kenya] by pouring in money." Onsando went further and said that KMYWO would not accept funding from "foreign countries who wanted to buy the country."[190] She vowed that KMYWO would "never accept any aid which might later be used to fight the Nyayo government."[191] In a lengthy speech to women in Mombasa, Onsando stated,

> Those who think they can use women for their own selfish ends are cheating themselves. The few disgruntled Kenyans and foreign donors who may think of using women in this country to carry out their devilish thoughts will never succeed. . . . We [KMYWO] shall play a big role in defending the country's peace and stability under President Moi's leadership . . . the foreigners who dream of buying us with money to destabilize the government will feel ashamed of even approaching us.[192]

Onsando was not specific in naming the foreigners or the countries to which she was referring, as the very foreign countries who were pressuring Moi and KANU to change were the same ones currently funding MYWO projects—primarily, the United States, through USAID which was funding at least 79 percent of all KMYWO's operations, and Germany, through the Konrad Adenauer Foundation.

KMYWO was apparently redefining its relationship with its foreign donors, calling for the conditional end to its partnership with them if their agendas included human rights, accountability and pluralism, while seemingly consolidating its marriage to KANU. As an alternative to foreign donor funding, Onsando called for women's groups to "search for self-reliance." This, however, was only rhetoric.

Many people and institutions, Kenyan and foreign, were calling for an end to the one-party state which had existed *de facto* since 1969 and *de jure* since 1982. They wanted a multiparty state and new elections. As Onsando had challenged foreign donors, she also challenged those who demanded change.

Moi had succeeded Kenyatta after his death in 1978 and had created a very closed and increasingly repressive authoritarian environment in Kenya. His administration was characterized by a culture of lies, manipulations, contradictions, embezzlements, corruption, mismanagement, human rights violations and political thuggery, of which KANU's behavior was symptomatic. Even the ethnicity of Moi's vice president and Minister of Finance, Professor George Saitoti, who is really a Kikuyu from Muranga District, Central Province pretending to be a Maasai, was lied about. This lie was promoted throughout Kenya with Saitoti's knowledge and consent, and probably even his orchestration, to maintain control and political hegemony through trickery.[193] As Nyayo literally means "footsteps," Moi had followed in the autocratic footsteps of his predecessor Kenyatta, although many would beg to differ.

In an edition of *Finance* appropriately entitled "Moi Time To Go," teachers, trade unionists, industrial workers, farmers, university students, doctors, the media, lawyers, priests and churches, former university professors ousted by Moi, Kenyan students abroad, the unemployed, *jua kali* artisans, politicians, young, old, cooperatives and even civil servants and former KANU rank-and-file members as well as KANU top-ranking members, all called for the end of the Moi regime.[194] Among these were outspoken women, including Mrs Wambui Otieno, widow of professor and advocate S.M. Otieno. Otieno had caused a stir in Kenyan politics when she fought an extended battle with the Luo Umira Kager clan for her rights as a widow, during which time women's organizations did not come to her assistance.[195] Otieno had been the one who publicly exposed Saitoti and challenged his vice presidential seat.[196] She and the groups aforementioned wanted political and economic changes and a new leader and government.

Many foreign donors also wanted changes and tied political and economic reforms to the continuing flow of aid to Kenya. They included: the Consultative Group chaired by the World Bank and comprising representatives from Canada, Denmark, Finland, France, Germany, Italy, Japan, the Netherlands, Sweden, Switzerland, United Kingdom, United States; the African Development Bank; the Commission of European Communities; the European Investment Bank; the International Monetary Fund; and the United Nations Development Program.[197]

Onsando and her national and district leadership campaigned heavily against those calling for reform. At the district level, Kirinyaga women, led by KMYWO District chair Fatuma Mohammed, announced on several occasions that they supported a one-party state.[198] In exchange, they were given a Ksh 300,000 donation by Moi, reminiscent of the exchanges between the colonial state and MYWO 30 years previously.[199] Moreover, women were advised by the KANU district officer to "avoid disgruntled elements wishing to plunge Kenya into chaos."[200] Shortly thereafter, KMYWO national office pledged to back the government against anti-government elements in the country.[201] Rallying their numbers, 42 district chairs of KMYWO, in a meeting chaired by Mrs Grace Ogot, MP and Assistant Minister of Culture and Social Services, resolved that their districts would hold a demonstration against advocates for a plural political system.[202] Nearly one year after this resolution, Aringo, then KANU's national chair, in response to the intensification of citizen's demands for reforms and multipartyism, urged women to "counter the campaigns against the Kenyan government by its critics and foreign governments and organizations."[203] Onsando, as the dutiful partner to KANU, responded on behalf of KMYWO. In an attempt to prevent the inevitable, she stated: "We have full confidence in the President, the ruling party, and the leadership of the country, we would like the multi-party advocates to know that they have no support from the women of this country."[204] KMYWO also threatened to sue the Law Society of Kenya for challenging the prerogative of the president in their request to dissolve the Parliament.[205]

Moi and the KANU government also used KMYWO to challenge political activists who held them accountable for their wrongdoing. Chief among them and their strongest female opponent was Professor Wangari Maathai, environmentalist and former wife of Lang'ata KANU MP Mwangi Maathai. Professor Maathai is the former chair of NCWK. Maathai is currently the coordinator of the Green Belt Movement (GBM) in Kenya and is an internationally acclaimed activist for human rights, environmental conservation, and social justice. The GBM had been started as a project by NCWK in the late 1970s. KANU's use of KMYWO to wrestle its opponents is obvious in the following example.

Maathai had challenged the government on its proposal to build a triple-tower 60-story building in the middle of Uhuru (Freedom) Park in Nairobi. It was planned to be the most prestigious piece of architecture in Africa, to be built by the Kenya Times Media Trust (KTMT) corporation, jointly owned by Robert Maxwell, international industrialist, and Aringo, KANU national chair. The building was to serve as, among other things, KANU party offices, and the entire cost of US $200 million was being borrowed for its construction.[206] Maathai opposed the construction on the grounds that a green belt of land measuring 1.5 acres would have been taken by KTMT for another unnecessary high-rise, and also because a national site commemorating Kenya's struggle for independence would have been destroyed. In a heated battle with scores of KANU heavyweight MPs who pressed for the building of the tower, Maathai argued:

> From Uhuru Park, one can see the magnificent Nyayo monument which is the symbol of peaceful transition. When the children of Nairobi walk through Uhuru Park, they are able to appreciate the symbols of victory as the children of Mombasa walking through Fort Jesus are able to recapture the oppressive and inhuman nature of slavery. . . . A Member of Parliament is not a deity.[207]

MPs argued that the high-rise should have been constructed in the name of "modernization." Maathai opposed them, linking modernization to environmental degradation, arguing that "Uhuru Park is the 'only green spot in the city which was fast developing into a concrete jungle'."[208] Not able to defeat Maathai by sound reason or logic, as they had none, the KANU men resorted to personal mud-slinging and degradation. They publicly misogynized Maathai, calling her "ignorant" and "a frustrated divorcee," plotted to "curse" her, and made references to her "anatomy below the line."[209] In spite of this, Maathai was relentless. KANU also used KMYWO to chastise

Maathai as they had chastised foreign donors and multiparty proponents. Maathai did not bend.

KMYWO "flayed" Maathai for her opposition to the KTMT complex. Spearheaded by women from Kilifi District, Coast Province, chair Mrs Beatrice Charo and Malindi Division KMYWO chair Mary Chizi called a press conference to disassociate KMYWO from the GBM, arguing that the latter was against the government.[210] KMYWO reportedly also vandalized the premises the GBM had occupied for ten years—government-owned wooden premises on land belonging to the Nairobi Central Police Station.[211] Further, KMYWO women allegedly uprooted and destroyed billboards belonging to the GBM. In KMYWO's defense, Onsando said that "maybe the women (KMYWO) had been angered by the movement and had acted to vent their frustrations."[212]

This was not the first time KMYWO was in confrontation with Maathai. KMYWO probably used this opportunity to get revenge for the previous two incidents in which Maathai blocked their plans against her. The first was in 1981, when "MYWO decided to pull out of NCWK [a coordinating body of women's organizations] after attempts by prominent members of MYWO failed to unseat [her as] chair."[213] Kiano, who was then chair of MYWO, and Maathai are from the same home district of Nyeri and they were political and personal rivals—Kiano on the side of the government, and Maathai on the side of principle confronting the government when it infringed on citizens' and environmental rights.

KMYWO and Maathai had also butted heads in 1984–5 when Shitakha tried to unseat Maathai as NCWK chair again. "Insiders saw the move as MYWO's bid to take over the activities of the council."[214]

KMYWO was impelled to oust and discredit Maathai because, over time, she had managed to cultivate an autonomous women's movement, separate and apart from MYWO, taking stands on women's issues, human rights and the environment against the KANU government; and she had won the battles consistently. Maathai had done what MYWO, with a reported membership of over 1.5 million women, had not. She had also empowered rural women in Kenya through the GBM tree-planting program.[215]

Maathai had not only held her ground against (K)MYWO, knocking them down each time, but had also won the battle against KANU MPs, despite their mobilization of KMYWO at the national level and its affiliate groups at sub-national levels to attack her. The KTMT proposed 60-storey building was not constructed, and Uhuru Park was saved as a national site.[216]

The state attempted to deny its loss to Maathai. It took steps to "punish" her for her headstrong and ungovernable behavior, particularly since "Kenyan politicians are not receptive to criticism 'especially from a woman who challenges state decisions'."[217] The court kicked the GBM out of its government-owned premises; the registrar general of the KANU government, Mr Joseph King'arui, ordered Maathai to furnish audited accounts of the GBM for five years prior to the KTMT incident; and KANU party members, with the sanctioning and assistance of KMYWO, continued their verbal abuse of Maathai. Nassir claimed that Maathai was against the Moi government and demanded that measures should be taken against her. KANU's reactions were not surprising. As Nzomo noted,

> Those women [*sic*] organizations or individuals within them that have resisted state control and/or challenged the oppressive status quo, have in the past often come under heavy censure and harassment, while the acquiescent ones have been rewarded and accorded high official status.[218]

PARTY AND PARTNERSHIPS BEGIN TO FALL APART

In intraorganizational relations, KMYWO became involved in the rifts that had begun to form in KANU. Although KANU had ruled Kenya as a one-party state, there were substantial, though latent, ideological differences between members of the party. Those differences began to widen and were publicly aired with discussions of multipartyism and democracy in Kenya in mid-1991. Some KANU members began to question KMYWO's affiliation to KANU. KMYWO's CEO, Jane Kirui, called those KANU members "hypocrites and political opportunists with chameleon tendencies."[219]

KMYWO's leadership support was to President Moi, whose style of rulership was to surround himself with personal loyalists. As multipartyism demands had become more threatening, however, Moi had engaged in several cabinet shuffles to ensure his centrality of power and reinforce his control as president, since some KANU members had begun to waver. Moi and KANU—party and government—were in trouble; and by affiliation KMYWO was in trouble. KANU was questioning and changing its relationship to KMYWO, a relationship which it had created. For example, KANU headquarters was requesting clarification as to whether or not KMYWO members were automatically members of KANU.[220] In addition, KANU was pressuring KMYWO to repay the party over Ksh 1.3 million which it had spent on elections,[221] as well as to turn in monies from sales of KANU memberships. In Nyeri District, Karatina KMYWO chair Veronica Mairimu Mugo was even jailed for allegedly stealing Ksh 156,530 from her group.[222]

By early 1992 a substantial number of KANU members had defected from the party, reportedly over a hundred from the Ikolomani constituency in Kakamega District alone.[223] Some KANU members had been fired by Moi, notably Aringo, who was the KANU chairman and the Minister for Manpower Development and Employment.[224] Key members of KANU had declared their allegiance and memberships to opposition parties. Moreover, many former KANU members and Moi political allies took the leadership of opposition parties to which KANU members defected.[225] For instance, Mwai Kibaki, former vice president (1978–88) and Minister for Health, was Moi's chief opposition in the Democratic Party (DP) with ally John Keen, a former Assistant Minister of State in the Office of the President and nominated MP. Jaramogi Oginga Odinga, Kenya's first vice president (1964–6) under Kenyatta and KANU first vice president concurrently, took the leadership of Forum for the Restoration of Democracy (FORD) Kenya. Martin Shikuku, former Moi ally in KADU and former Assistant Minister of Home Affairs, Minister in the Office of the President and Assistant Minister of Livestock, was a founding member of FORD Kenya. Kenneth Matiba, former Minister for Culture and Social Services, permanent secretary in various ministries and MP for Kiharu, and Moi's toughest opposition, took the leadership for FORD Asili. Kibaki and Keen had defected from KANU. In 1982 Odinga had been expelled by Moi from KANU. Shikuku had been dismissed by Moi in 1984, and Matiba had resigned from the Moi Nyayo government, and had been detained by Moi and expelled from KANU.[226]

Concomitantly, there was also growing confusion in KMYWO and foreign donor interorganizational relations. KMYWO tried frantically to woo itself back to the donor community, despite the allegations that it had made against donors while defending Moi and KANU. In a twist of fate, KMYWO became more dependent on its interorganizational linkages to foreign donors as its ties to KANU began to unravel. KMYWO took desperate measures to save itself from disintegration with what seemed to be a dying party.

After having denied that KMYWO had lost foreign donor support,[227] on 7 December 1991 Onsando capitulated and announced that three of the organization's programs had ceased to operate due to foreign donor withdrawals. She identified the withdrawing donors as NORAD, Marttaliitto of Finland and the Konrad Adenauer Foundation of Germany. Onsando expressed concern for the job security of KMYWO members who were employees of these programs, paid by donors.[228] Onsando made no reference to the loss of foreign donor support being connected to KMYWO's affiliation to KANU.

Other prominent politicians, however, led by Mrs Agnes Ndetei, MP for Kibwezi and one of very few women in Parliament, made the connection clear and public, and requested that KANU end its affiliation with KMYWO on the grounds that foreign donors "were unwilling to assist the organization because of its links to KANU." Ndetei's request was seconded by MP Peter Okondo from Bunyala.[229]

On 11 December 1991, Mrs Grace Ogot, MP for Gem and Assistant Minister for Culture and Social Services, announced to Parliament that "KANU and KMYWO have signed an agreement [to disaffiliate] to avoid confusion about relations with foreign donors."[230] This was the first time that KANU had acknowledged that a problem with foreign donors had arisen from its affiliation of KMYWO.[231] At this time, KMYWO would revert to the use of its old name, MYWO.

On that same day, Onsando retracted her statement of one week earlier. She said that foreign donor assistance to KMYWO had not been cut. She argued that foreign donors had simply handed projects over to KMYWO to be self-sustained. Onsando further said that new agreements had been made with new international donors whom she did not identify.[232] The evidence surrounding the withdrawals of foreign donors referred to by Onsando in her first statement did not support her story.

NORAD had been kicked out of Kenya because of diplomatic disagreements between Kenya and Norway over Koigi wa Wamwere, a political dissident and former MP, who had sought and been granted asylum in Norway. Norway had attempted to intervene in discussions regarding Wamwere, who was accused of having clandestinely returned to Kenya carrying contraband (guns and grenades), for which he was charged with treason. Moi kicked NORAD out of Kenya for interfering in the sovereign affairs of the state. As a consequence, some program areas of MCH/FP which had been covered by NORAD were no longer covered.[233] Marttaliitto had not renewed its funding for the Nutrition program because of KMYWO's affiliation with KANU, under orders from the Finnish government.[234] The Konrad Adenauer Foundation pulled out in 1992, after two three-year contracts with MYWO funding its LD program. It pulled out because of MYWO's and KMYWO's gross inefficiency, mismanagement and overall non-compliance with the foundation's requests for necessary financial reports and evaluations.[235]

Despite KANU's official disaffiliation with KMYWO, unofficially their relationship remains the same. MYWO continues to support the state and echo its views.[236] Disaffiliation did not sever the loyalty that MYWO's current national leadership—Onsando and her national executive—and their patron, Kiano, has for KANU and for Moi. Onsando publicly announced that she planned to remain a part of KANU, despite the disaffiliation.[237] Her national executive is sticking with her. Kiano demonstrated her quiet but solid support for KANU during its process of affiliating MYWO to the party, and this continues. She remains its watchful and influential patron. Moreover, the current MYWO leadership, elected as KMYWO leadership, is holding on to power, postponing until 1995 elections which should have occurred in 1992, according to the (K)MYWO constitution.

Disaffiliation of KMYWO from KANU was merely a symbolic act for foreign donors to again pour money and other assistance into MYWO and Kenya. It was a desperate measure to reactivate interorganizational trilateral relations as they had existed when Kiano was chair and when MYWO was at least a quasi-NGO, somewhat autonomous, yet unofficially but closely linked to the state. Nothing has really changed—except the rhetoric of (K)MYWO leadership. They are the real chameleons—not at all committed to representing Kenyan women's grassroots interests. Kiano remains, though less so today, the linchpin between MYWO and KANU and a fading link to foreign donors.

4

A CHANGING RESEARCH METHODOLOGY AMID POLITICAL VOLATILITY, ENVIRONMENTAL UNCERTAINTY AND A CULTURE OF FEAR AND SILENCE

This chapter presents the research strategy that was used in the study for collecting data. This strategy evolved through attempts to grapple with questions of resource dependency, assistance, autonomy, balance of power in IOR, indigenous development, gender and self-reliance as they relate to MYWO and the organizations in its relational environment. Furthermore, it engages these issues against the backdrop of heightened political unrest and foreign aid uncertainty during the period of my research, August 1991 to August 1992.

The chapter is divided into four parts. Part I highlights the major political and economic changes in MYWO's relational environment which took place just prior to or during the period of my research and which had direct effects on the research strategy of this study. A discussion of these changes is important since they contextualize the relational environment of development cooperation within which MYWO engaged organizationally and politically with its partners in development. Part II demonstrates the unsuitability, partly based on MYWO's compromised autonomy and its dubious status as an NGO, of the research hypotheses as initially formulated. It further shows the necessity for the reconceptualization of the strategy for investigating the development cooperative partnership between MYWO, foreign donors and the Kenyan government. This strategy was redesigned in the field from hypotheses testing to a more exploratory inquiry. Part III of this chapter discusses the study instrumentation, as well as provides a description of the study participants, archival research and events observation. Part IV discusses the benefits and limitations of the actual research strategy utilized.

PART I: POLITICAL AND ECONOMIC CHANGES IN KENYA, AUGUST 1991–AUGUST 1992

The call for multipartyism

The one-party state of the Kenyan African National Union (KANU) had existed *de facto* since 1969 and *de jure* since 1982. Kenya had become independent on 12 December 1963 and Jomo Kenyatta, the head of KANU, had become Kenya's first president. Other political parties thereafter disappeared. The minority party, Kenya African Democratic Union (KADU), of which Moi was a key member of the leadership, dissolved itself in 1964 and joined KANU. A small leftist opposition party, the Kenya People's Union (KPU), surfaced in 1966 led by Oginga Odinga, who had been Kenyatta's vice president 1964–6. In 1969, KPU was banned and KANU became the sole ruling party, thus making Kenya a *de facto* one-party state. In 1982, under President Moi, the Kenyan constitution was amended, making Kenya a *de jure* one-party state.

Almost 30 years after Kenya's independence (1991–2), Kenyans were demanding an end to the one-party state and the autocratic rule of President Daniel arap Moi. They were calling for multiparty politics and elections for a new president. Amid resistance and repression from KANU, other parties began to form. Tensions heightened between the citizens and the state. One of Moi's and KANU's tactics to solidify their hold on state power had been to coopt or dissolve national non-governmental organizations. MYWO had been coopted, as discussed in Chapter 3.

KANU's attempt to consolidate the state had a direct impact on access to interviewees, thus affecting my research strategy. Many Kenyans refused to talk to researchers because of legitimate fears of reprisals from the Kenyan government. Moreover, they did not know whom they could trust. Of those who did agree to interviews, many refused to talk about MYWO, particularly its elected officials at the national level and their relationship to KANU and the state. MYWO, as a state organization, had become a mouthpiece in defense of the state, and it was not amenable to being researched.

Foreign donor withdrawals

During the last quarter of 1991, donors began to threaten withdrawals of foreign assistance to Kenya. Foreign donors posed withdrawal threats as conditions tied to human rights violations, corruption and political and economic reforms. The United Kingdom, Canada and Germany were among the first nations to tie assistance to human rights violations, stating that they could not subsidize repression. Moreover, NGOs from 14 Commonwealth African nations at the Commonwealth Heads of State meeting in Zimbabwe in October 1991 demanded the imposition of sanctions on countries violating human rights. They spotlighted Kenya.[1]

Corruption by Kenyan officials was a major factor causing Western nations to cancel aid. Denmark, Britain and the US argued that they were no longer confident that aid was reaching the people for whom it was intended. Northern NGOs also expressed concern, and linked their concern to the flow of NGO assistance to Kenya. To demonstrate this corruption, the American Ambassador in Kenya, Smith Hempstone, reported in a "get rich now cable" in 1990 that four Kenyan government officials and three relatives of President Moi had "greatly enriched themselves through their public office or access to officials." The embassy listed the names as President Daniel T. arap Moi, Minister of Energy Nicholas Biwott, Vice President George Saitoti, and Permanent Secretary in the President's Office for Internal Security, Hezekiah Oyugi. Biwott, like Moi, is a Kalenjin. He is considered to be Moi's closest political confidant and is rumored to be Moi's nephew. Biwott's estimated wealth is in the hundreds of millions of dollars, while his salary as energy minister was approximately $750.00 per month in 1991. In the middle of the donor freeze on foreign aid to Kenya, Biwott had successfully obtained a Kenya High Court injunction blocking New York's Citibank from collecting over US $14 million worth of loans to his companies, most of which had been secured through credit to Biwott's Swiss bank account.[2]

To further speak to the corruption allegations, President Moi is also known to have benefited from partnerships and payoffs in countless business ventures. Moi is believed to be a partner in an oil refinery in Puerto Rico with an Israeli businessman; he is also believed to have a share in a Nairobi casino and a hotel in the Maasai Mara game park; he is believed to be a member of the Dolphin Club, and one of the primary owners of the Nairobi Trade Bank. It is also alleged that Moi received a "seven figure" commission from the European Consortium of Airbus Industries from which the Kenyan government purchased two planes. Two Kenyan businessmen are believed to be front men for Moi. Moi provides the business opportunities and the front men provide the "cut." Newly emergent Asian business tycoon Ketan Somaia is reportedly the main front man. Somaia is regarded by elder Asian businessmen as "an 'overzealous favor seeker' who enriched himself as a proxy of political godfathers."[3]

It appears that it has become practice and even expectation that Kenyan politician–businessmen amass large sums by siphoning aid, demanding kickbacks and even shares and controlling interest in companies that start businesses in Kenya. Though Kenyatta's acquisitions in land, precious gems, ivory and casinos from 1963–78 were great, they pale in significance compared to what economists contend Moi and some of his associates have acquired.[4] Some of MYWO's leadership have even tried to get their cut by endearing themselves to Moi and his associates through KANU.

The lack of political and economic reform was also a major reason for suspending aid to Kenya, as mentioned in Chapter 3. Donor nations were calling for democratization of the political structure and liberalization of the economy. Among the more specific demands of the donors were the adoption of a multiparty democracy and structural adjustment programs (SAPs) including the privatization of the public sector, particularly the sale or liquidation of parastatals, a reduction of the budget deficit and a reduction in the over-employment in the Civil Service, whose wage bill accounted for 50 percent of the total government expenditure. As no signs of progress sufficient for donors were shown in these areas, in November 1991 donor governments suspended their annual practice of pledging aid to Kenya. They also suspended most "budgetary facilities and balance of payments support."[5]

The Kenya government's reaction to foreign donors was that Kenya was being abused by the West, and that the sovereignty of the Kenyan state was being undermined. For example, Moi made appearances to the *wananchi* (citizens), inciting anti-Western sentiments and announcing that he would not take the abuse, while he simultaneously and hypocritically made promises of change to donors. In support of Moi's anti-Western stance, the Secretary General of the Central Organization of Trade Unions of Kenya (COTU), Joseph Mugalla, stated that the World Bank and the IMF wanted to "enslave" Kenyans by creating "mass unemployment and depreciating wages." The Secretary General of the Organization of African Unity (OAU), Dr Salim Ahmed Salim, agreed with the sentiments of Moi, declaring that aid conditionality amounted to "blackmail" of African countries.[6] Though both sides had some justification for the positions they took—the donors were correct in that political and economic reforms were overdue in Kenya, and the Kenyan government was correct in that SAPs had actually worsened instead of improved structural inequities in the economies—there were few signs that the Moi government was willing to make changes, and few signs that foreign donors would renege on their demands.

Many of the Kenyan government's reactions to donor withdrawals, which were played out in anti-Western sentiments, roused Kenyans' suspicion of foreigners. This had an impact on my research design because it affected the willingness of Kenyans to be interviewed. Many simply would not talk. Moreover, information that circulated and implicated high-level politicians in corruption had an impact on my research design, as well as fostered the culture of fear and silence. Many who agreed to interviews cautioned me against who and what topics I should not broach; commonly mentioned were the Dolphin Group and the Trade Bank. The political and research climate at this time was particularly tense. MYWO's focus was on defending KANU and its corrupt practices. It was not open to being interviewed, as it could also be exposed.

The Ouko trial

While the interorganizational reconfiguration, negotiations and backbiting was going on between foreign donors, the Kenyan state and MYWO, a Commission of Inquiry was simultaneously conducting a judicial probe into the events surrounding the February 1990 murder of the Minister of Foreign Affairs and International Cooperation, Dr Robert Ouko. This inquiry was being held at the time I began my research in Kenya. It had begun on 16 October 1990 and continued until 26 November 1991, when it was abruptly halted by Moi. A total of 173 witnesses had testified, including Inspector Graham Dennis and Detective Superintendent John Troon of Scotland Yard. Allegations were made and suspects were named concerning several incidents which could have been the precipitating causes of Ouko's murder. These incidents (as they will be discussed in the following pages) had a critical impact on the research climate during the time I was in the field.

One incident which many interviewees raised involved Biwott, who had made a sarcastic referral to Ouko as the president of Kenya in January 1990, when a delegation of Kenyan officials had traveled to the US to meet President George Bush to request financial assistance for the construction of the Kenya Times Media Trust Building. Biwott was said to be incensed by the "preferential treatment" he perceived Ouko was receiving from US officials. Moi was reportedly angered and jealous of the manner in which Ouko was being received. Magendo rumors allege that this incident prompted Ouko's elimination. Ouko was killed one month later.[7]

A second incident involved the revival of a Kisumu molasses plant which had folded in 1982 without ever having been opened, though it was 95 percent complete. Ouko was attempting to rehabilitate the plant. He was said to have been negotiating with several investors. Biwott and Saitoti had different plans, however. They were negotiating with a rival group of investors, against the plans of Ouko. Reportedly, Biwott and Saitoti worked together to "frustrate Ouko." Saitoti would not sign the necessary papers for Ouko to begin the project, and Biwott simply did not like Ouko's choice of investors. Furthermore, Biwott wanted a commission for the project from investors as a payoff. During the inquiry it was reported that Biwott had received a bribe of Ksh 85,000,000 from the Asea Brown Boverie, a foreign electrical equipment firm.[8] This leads to the third incident.

The third incident was Ouko's completion of an anti-corruption report. Ouko, who reportedly never accepted a bribe, spoke out strongly and loudly against high-level corruption. He had compiled a report on corruption which he was to submit to Moi. Many politicians, including Biwott, Saitoti and Oyugi disliked Ouko's stand in this regard because many of them were vulnerable to being exposed.[9]

A fourth incident was the rivalry of Ouko and Mr Job Omino for the

Kisumu Town parliamentary seat in the 1988 general election. As the inquiry probed into this rivalry, Ouko's former campaign manager, Mr John Eric Reru, named "powerful individuals" who stood against Ouko for fear that Ouko might be appointed vice president of Kenya. Reru named Biwott, Oyugi, the former Nandi District KANU branch chair Mark Too, the then Minister for Local Government and Kakamega District KANU branch chair, Moses Mudavadi, and one more person who the inquiry deemed could not be publicly identified.[10] They encouraged the rivalry between Ouko and Omino.

A fifth incident involved an alleged affair Ouko had with the wife of Nakuru District Commissioner Mr Jonah Anguka. *Magendo* sentiment was that the rumor of the "affair" was a diversion, and that Anguka was the "fall guy" used to deflect suspicion from the real murderers. Attention was being placed on Anguka as a jealous husband who murdered Ouko in a fit of rage.

Scotland Yard's Detective Superintendent John Troon had been charged with the investigation of Ouko's murder. In his report, Troon named Biwott and Oyugi, who had by then become *former* minister and *former* permanent secretary respectively, as prime suspects. Anguka and the former Nyanza provincial commissioner Julius Kobia were identified as "suspects on the periphery." Troon's report called for further questioning and investigation into the matter, but Moi immediately dissolved the inquiry upon the release of the report. This dissolution came as a surprise to the judges holding the inquiry. Shortly thereafter, Biwott, Oyugi, Anguka and the Ouko family lawyer, George Oraro, were arrested. The *magendo* rumored that the arrests occurred abruptly because Oyugi asked to testify to the Commission. Two weeks after the arrests, Biwott and Oyugi were released for "lack of evidence," and Anguka was charged with murder.[11]

Dr Ooka Ombaka, the lawyer who represented Ouko's Ominde Society during the inquiry, protested against Moi's dissolution of the Commission, calling it unconstitutional. He further stated that from the evidence presented in the inquiry "Moi knew the identity of Ouko's killers." Ombaka was picked up for questioning by the police.[12] He was also linked to Dr Wangari Maathai, who is his client (and who was discussed in Chapter 3).

All of these events related to the Ouko trial stirred public fear, and this inevitably had an impact on my research design. I was not able to interview anyone from the Ministry of Energy or the Ministry for Foreign Affairs and International Cooperation—that is, in either Biwott's and Ouko's ministries. They were very suspicious of my motives as a researcher, and wary of journalists or spies who might pose as researchers. MYWO made no public statements about the Ouko trial.

Citizens' reactions and demonstrations against Moi and the KANU government

During the fieldwork, several institutions and segments of Kenyan society protested against Moi and the KANU government. These institutions included the Law Society of Kenya (LSK), of which Paul Muite was chair and Dr Willy Mutunga was vice chair. Since the 1980s, LSK had been one of few active opposition forces against the government and for this they have received international recognition and support, having received the International Human Rights Award from the American Bar Association in October 1991. The National Christian Council of Kenya (NCCK), of which Sam Kobia is head, has also been a force of opposition against the repression of the Moi government. The Kenya Professional and Business Women's Club (KPBWC), chaired by Beth Mugo, not known for its particularly political posture, also took a staunch opposition stance against the government. It requested that the government resign because of its instigation of and involvement in "ethnic clashes." The citizens' demonstration that drew the most national and international press attention and public participation, however, was the Mothers' Hunger Strike.[13] The press, particularly the magazines *Society*, *Finance*, *Weekly Review* and *Nairobi Law Monthly*, remained critical of the government, particularly of Moi. Over time, they became overtly anti-Moi, reflecting widespread anti-Moi public sentiment. The government claimed that these magazines had "'godfathers or foreign masters funding them."[14] The three daily newspapers also provided very current and detailed information of the issues. During the period of my research the press became more open as it distanced itself from the influence and censorship of the Kenyan government.

In addition, virtually all of the persons who took stands against the government were either constantly harassed or running underground because of fear of harassment or worse. Non-Kenyan associates of these persons were also targets of suspicion. In November 1991, an American student working with members of the LSK was attacked, picked up for questioning by the Special Branch and then told to leave Kenya and taken by escort to the airport. This invariably had an impact on my research strategy, especially because of the political nature of the politics of development cooperation. I was asking questions at a time of heightened tension and suspicion of government and foreigners.

The formation of new political parties in Kenya and the repeal of Section 2(a) of the Kenyan Constitution

Despite Moi's reluctance to open up the Kenyan political structure and reintroduce multiparty politics, the fervor of multipartyism had gripped the Kenyan citizenry by mid-1991. Moreover, the donor community was exerting more and more pressure, including tying aid to democratization. Moi continued allegations that "foreigners" were dictators of African political systems for which Africa was not ready; that foreigners were imposing their will on Africa; and that African systems of government which now existed emanated from African political traditions.[15] MYWO supported Moi in making these allegations, yet these allegations did not stop the inevitable. At the scheduled consultative donor meeting in Paris in November 1991, Vice President Saitoti humbly presented a privatization plan for Kenya and further stated that Kenya was fighting corruption and was not against a pluralist democracy. The participants in the conference agreed that "Kenya's economy has been highly dependent on external assistance and . . . continued and significant aid flows are needed to support Kenya's economic development."[16]

On 6 December 1991, Moi finally recommended that the delegates at the Special Delegate Conference at Kasarani approve an amendment to repeal Section 2(a) of the Kenyan Constitution which made Kenya a *de jure* one-party state on 9 June 1982, thus officially reintroducing multipartyism in Kenya.[17]

That very day, Forum for the Restoration of Democracy (FORD) officially launched its party, followed at later dates by the Democratic Party (DP), the Social Democratic Party (SDP), the Islamic Party of Kenya (IPK), Democratic Movement (DEMO) and other smaller parties. On 8 June 1992 voters began to register for the general elections which were held in December 1992 amid several postponements.[18]

There was suspicion that foreigners in Kenya were inciting the creation of parties to destabilize Kenya, thus affecting my research. Onsando, the chair of MYWO, geared her speeches to women against foreign "meddlers" and "imperialists."

Ethnic clashes

During the last quarter of 1991, ethnic clashes between neighbors had erupted, displacing and killing thousands of Kenyans. This inevitably had an impact on my research. The clashes began in October between the Kalenjin and the Luo, reportedly for control of land and farms in the Rift Valley and Nyanza Provinces, which during the colonial period had been designated as "White Highlands." The clashes spread thereafter and included other ethnic groups as well: Kisiis, Luhyias, Kikuyus and Maasais. Questions began to circulate about other reasons for the clashes, extending beyond land questions. On the surface, the clashes seemed to be more about one ethnic group fighting another, not only in the rural areas, but in Parliament as well. Two Kalenjin politicians, Biwott and Kipkalia Kones, and one Maasai politician, William ole Ntimama, told their constituencies to "arm themselves and drive out 'foreigners' " (other Kenyans) from the

Rift Valley Province. Moi himself challenged the collective manhood of his Kalenjin clansmen, revealing his ethnocentrism and misogyny. Moi stated "that he would consider them [Kalenjin men] 'women' if they let the monster [multiparty] set foot in the province."[19]

Many Kenyans insightfully pointed out that the clashes were not really ethnic, as the instigators would have the international community believe. They were instead political and economic clashes, not perpetrated by all Kalenjin, but instead by a small group of Kipsigis Kalenjin led by Biwott and Moi, who had inter-ethnic co-conspirators such as Saitoti and some Maasais. Kipsigis Biwott and Moi seemed to have more in common with Kikuyu Saitoti (who poses as a Maasai), both politically and financially, than they do with their intra-ethnic clansmen and rivals, the Nandi Kalenjin. The Nandi have historically been more powerful than the Kipsigis. As evidence of their involvement in ethnic clashes, in March 1992 Biwott was identified as the supplier of arrow-heads imported from North Korea to his army in Kalenjin territory. Customs officials at the Jomo Kenyatta International Airport confiscated the arrows en route from North Korea, and Biwott was subsequently questioned. In June–July 1992, Saitoti's farm in Molo was discovered to be a storage place for arms used in the so-called ethnic clashes.[20] All evidence pointed to top KANU party and state officials as the perpetrators of the "ethnic" clashes.[21]

It is believed that KANU stalwarts encouraged the clashes for several reasons: 1) to demonstrate to the international community that although section 2(a) of the constitution had been repealed, Kenya was not ready for multiparty democracy because "tribalism" was volatile and divisive; 2) to prevent the registration of voters for the general election because Kenyans were ready to oust Moi. The logic was that Kenyan potential voters would be too fearful of registering, hence Moi would hold on to power; 3) to bring about *Majimboism* (a federal system of government) allowing the Rift Valley Province, in which the Kalenjin live, to become an autonomous region; and 4) to deflect from the real problem—the determination of a small group of politicians-thieves-thugs to hold on to power without popular support. MYWO supported this small corrupt group. MYWO did not take a stance against the clashes, either. Ethnic clashes very seriously affected my research strategy. It made many project areas off-limits to my investigation. For instance, I stayed clear of the Rift Valley Province where most of the clashes were occurring. I was also caught in a clash in Kipkarren, Kakamega District, Western Province.[22]

These political and economic events invariably had a significant impact on my ability to carry out my research design as I had proposed. They called for insight, flexibility and caution in all situations. I was never really certain of the stability of the situation at hand, or the stability of the country from one day to the next. During the period of my research, the Kenyan political environment was extremely tense and potentially volatile in virtually all respects. Kenyans were challenging and reshaping the Kenyan political culture, as well as the post-colonial "culture of politics."[23] These events and my research experiences are the backdrop against which I contextualize MYWO's actions in the politics of development cooperation with its partners—foreign donors and the Kenyan government.

From the calls for multipartyism to ethnic clashes, it becomes understandable how the repression of Kenyan citizens by the Moi government had created a culture of fear and silence in Kenya. When I arrived there in August 1991, most people would not talk about the state, the government, Moi or MYWO to each other, much less to a researcher. One could not even mention Moi's name in public. There was widespread suspicion of spies among the Kenyan population, and of researchers from abroad.

Women, particularly, were muted by this culture of fear and silence as they feared repercussions from the government, as well as from the men in their families, for "conspiring with foreigners" against the government. This culture of fear and silence permeated women's groups, and affected their interaction with me. It may have been a reason MYWO officials would not talk with me, and a reason why many women would not talk about MYWO.

PART II: RECONCEPTUALIZING AND REDESIGNING THE RESEARCH STRATEGY

Initially, before the fieldwork and before the eruption of the political and economic volatility in Kenya in 1991–2, this study assumed a causal relationship between financial and technical assistance (the Independent Variables—IVs) provided by foreign donors and the Kenyan government to MYWO, MYWO autonomy (Intervening Variable), and success or failure of MYWO programs and projects (Dependent Variables—DVs). As stated in Chapter 1, the major hypotheses were:

1 The greater the financial and technical assistance MYWO receives from foreign donors and the Kenyan government, the less autonomy MYWO will have to formulate and implement its indigenous agenda, hence the less successful MYWO development projects and programs will be.
2 The less the financial and technical assistance MYWO receives from foreign donors and the Kenyan government, the more autonomy MYWO will have, hence the more successful MYWO development projects and programs will be.

For more details about 1) the initial hypotheses, 2) the key variables and their operational definitions, and 3) a graphic illustration of the initial hypotheses see Appendices AI–AIII. Very early in the fieldwork, however, amid political volatility and environmental uncertainty, as well as the culture of fear and silence which had been created by real and threatened government repression, tentative observations of the interactions, negotiations and politics of the development cooperative partnership under investigation suggested that the hypotheses as initially proposed were incapable, in their simplicity and linearity, of capturing the complex interaction as well as the interplay of various political forces that influenced and shaped MYWO's relational environment. These forces included, but were not limited to, MYWO's foreign donors and the Kenyan government. There were other confounding and extraneous variables that made linkages between cause and effect extremely difficult, and perhaps even impossible to isolate. For instance, the Kenyan citizens' demands for changes in the "culture of politics" of the Moi Nyayo government and the international pressures on the Kenyan government, spearheaded and stringently applied by the World Bank and the IMF to pluralize politically and liberalize economically, permeated the relational environment of MYWO and its partners in development, affecting their politics of development cooperation. In addition, the 1987 cooptation of MYWO by KANU further confounded the research strategy because the cooptation made it fruitless to consider MYWO's autonomy in relation to its partners in development. MYWO's relative organizational autonomy after the resignation of Jane Kiano as chair in 1984 had been compromised at best, and after the removal of Theresa Shitakha as chair in 1985 the autonomy of MYWO was virtually nil. Thus, MYWO's status as an NGO was dubious. Although the organization continued to claim that it was an NGO and presented itself as such, particularly to foreign donors, MYWO by affiliation to KANU was inextricably tied to the ruling party and the state. It had become, for all intents and purposes, an organ of the state. KANU was its patron.

MYWO's autonomy was problematic yet particularly intriguing for this research, since autonomy is not a fixed and tangible quality. Autonomy is instead a fluid, intangible and variable concept. One of the problems with the NGO literature at present is that references to NGO autonomy are frequently made, yet scholars have yet to operationalize a working definition for the concept. Moreover, a distinctive quality and the very essence of an NGO (according to its most comprehensive and useful definition to date as discussed in Chapter 2) is that it is "autonomous" from government, yet the factors that constitute autonomy have not been clearly defined. For instance, this NGO definition states that an NGO should not receive "substantial" contributions from its home government. There is not consensus or even discussion, however, of what constitutes "substantial." Audrey Wipper, in the 1970s, raised questions with regard to MYWO autonomy. These questions continue to pose challenges, particularly in this era when NGOs are fast forming partnerships in development cooperation.

A final reason why the initial research strategy proved unsuitable in the field is that the types of data that were presumed to be accessible, such as the amounts of financial assistance and the degree of technical assistance that foreign donors and the Kenyan government provided to MYWO, were not available from the field. Donors (with the exception of Pathfinder who provided limited information) would not disclose the figures in interviews either.

Hence, for these reasons, the initial research strategy was reconsidered and reframed to reflect the complexity and circuity of the politics of development cooperation between MYWO and its development partners, as those politics play out in the field. The manner in which the reconceptualized research strategy reshaped the interview questions will be discussed in the next section of this chapter as well as in parts of Chapters 5 and 6.

PART III: STUDY INSTRUMENTATION—ORAL INTERVIEWS

This research was conducted in two phases: 1) in Kenya, from August 1991 to August 1992, and 2) at the Public Record Office (PRO) in England in August 1992. Some preliminary data had also been gathered at the PRO in July 1989, and some observations had been conducted in Kenya from July 1989 to October 1989.

In Kenya in 1991–2, the study instrumentation that was utilized was an oral interview survey method. Open-ended interview questions were drafted and used as a guide for the interviews, adapting the questions for the peculiarities of each individual situation, and allowing for deviations from questions so that interviewees could tell their stories in their own ways and share their perspectives on MYWO and the politics of development cooperation in Kenya. A detailed list of those questions is provided in Appendix B. All the responses were recorded by hand, during the time of the interview or as soon as possible thereafter. The interviews were conducted in Kiswahili and in English, determined by each individual research situation, and the preference of the interviewee(s). Sometimes both languages were used.

There were four specific interview populations: 1) local grassroots women, 2) MYWO officials and staff, 3) foreign donors, and 4) government bureaucrats. There also emerged a fifth set of interviewees that will be referred to as "other relevant persons." They will be discussed in greater detail later.

The questions asked of each group attempted to solicit general perceptions about issues of indigenization, assistance, autonomy, dependence, MYWO program and project successes or failures, and relations between and among partners in development.

The study participants: descriptions

During the period of my fieldwork in Kenya, I interviewed 106 research participants. Of this total, 24 were identified as women's groups, whose numbers of members varied (as discussed in the next section), 10 were past and present MYWO elected officials, 16 were MYWO staff persons, 16 were representatives of foreign donors, 19 were Kenyan government ministry personnel, and 21 were other relevant persons.

Women's groups

The 24 women's groups represent five of Kenya's seven provinces and the Nairobi area. (Nairobi is not incorporated in any of the seven provinces.) Of these groups, 15 were rural and 9 were semi-urban. One of the groups was started in 1958, 5 were started in the 1970s, 6 were started in the 1980s and 6 were started in the 1990s. Six groups gave no starting date or year. The groups ranged in size from 6 members to over 1,000 members. The larger size groups were usually coalition groups of most or all women's groups in a particular area. There were 6 coalition groups in this study. The reasons that these groups evolved as interviewees is discussed in Part IV of this chapter.

MYWO elected officials and staff

Four national elected officials were interviewed. Among them, 2 were national officers from the Kiano administration and 2 were national officers from the current Onsando administration. They were the national MYWO chair Mrs Wilkista Onsando and her assistant national treasurer Mrs Joan Mjomba. One provincial elected officer was interviewed. Two district chairs were interviewed. One divisional chair and one divisional treasurer were interviewed. One village chair was interviewed. Except for the two elected officials from the Kiano administration, all of the officials interviewed were elected to office during or after the MYWO 1989 elections.

Of the 16 interviewees, 9 were from the national headquarters. They included MYWO's chief executive officer (CEO), 4 national program managers, 2 assistant program managers, an archivist and a national accountant. From the district level, 2 accountants and 5 field workers were interviewed. All of the persons interviewed were MYWO personnel serving in the current Onsando administration.

Foreign donors

A total of 9 foreign donors, currently providing or having recently provided financial and/or technical assistance to MYWO, were interviewed. They included: an IGO, the World Bank; 2 GOs, the United States Agency for International Development (USAID) and the United States Peace Corps; an INGO donor, Pathfinder International; 4 NGOs, the Center for Population and Development Activities (CEDPA), the German Technical Agency (GTZ), the Konrad Adenauer Foundation (KAF) and Marttaliitto of Finland; and a business donor, Coca-Cola–Kenya. These donors are categorized here according to the organizational definitions which they provided. All provided financial and/or technical and/or material assistance to MYWO.

A total of 16 foreign donor representatives of the aforementioned donors were interviewed. A brief description of these foreign donors is provided in Appendix C. Findings with regard to their operating functions as NGOs, INGOS, GOs, IGOs and businesses, which sometimes conflict with their self-definitions, and their donors relationships to MYWO will be presented in detail in Chapter 5.

Kenyan government ministry bureaucrats

Nineteen persons were interviewed from various government ministries. One was from the Ministry of Livestock. Six were from the Ministry of Agriculture. Of these, five were from Home Economics Extension and the other was a Special Appointment for Irrigation Rehabilitation. Ten interviewees were from the Ministry of Culture and Social Services (MCSS), holding various positions within that ministry. Two interviewees were from the Ministry of Home Affairs and National Heritage. Ministry personnel are almost always university graduates. The interviewees were from four provinces and the city of Nairobi. Although no one was interviewed from the Ministry of Health, this ministry was involved with MYWO in the implementation of the MCH/FP program. A brief description of the ministries is provided in Appendix D.

The Ministry of Energy was to be included in the interviews. However, this ministry was undergoing major crises during the period of my fieldwork, and was thus not open to interviews.[24] No interviews could be conducted with the Ministry of Foreign Affairs and International Cooperation either, because the murder trial of its former minister, Dr Robert Ouko, was ongoing.

With regard to the role of Kenyan ministries in development, the Ministry of Health planned programs to ensure the health quality of the nation. The Ministries of Agriculture and Livestock Development provided technical experts and specialists to assist with programs and projects which were in keeping with the government's objectives. They provided skills and training to groups and individuals, particularly farmers. The MCSS provided various types of motivational support to groups engaged in community development activities. The Ministry of Home Affairs and National Heritage created the National Council for Population and Development (NCPD), under the Population Guidelines contained in Sessional Paper No. 4 of 1984, as the government arm to coordinate and monitor all programs geared toward family planning.[25]

Other relevant interviewees

Interviews and discussions were also conducted with a total of 21 other relevant persons. They included government officials such as district commissioners and district officers. Others included international NGOs not funding a MYWO project, such as Danish International Development Agency (DANIDA); women who were not members of MYWO; KANU women's league members, some of whom were members of women's groups; other grassroots organizations not affiliated with MYWO; political activists, academicians and professionals.

This group of "other relevant interviewees" emerged as an unintended group. It came about because all of the avenues which I had planned to pursue to reach interviewees were closed when I arrived in Kenya. The country was undergoing very tumultuous times, some of which were discussed in Chapter 3 and Part I of this chapter. For the first few months in the field in Kenya, it was impossible to get an interview with anyone associated with MYWO, especially MYWO at the national headquarters, and this was one of the main reasons the unintended group emerged.

Archival research

During the period of my fieldwork in Kenya from August 1991 to August 1992, I also conducted archival research at MYWO headquarters, the *Kenya Times*, the *Daily Nation*, the *Standard*, the MCSS Women's Bureau and the Kenya National Archives. Also in August 1992, I conducted archival research at the Public Record Office in England.

Events observation

Fieldwork in Kenya provided an opportunity for direct observation of events and activities, both planned and spontaneous, that had direct bearing on my research. I was able to observe the behavior of organizations and persons directly relevant to the research problem in attempts to unravel the politics of development cooperation. I was also able to observe events in their actual environmental context in which this research problem unfolded.

In conducting my observations, my role shifted between 1) participant-as-observer, at which times I participated with the groups I studied while I made clear that I was conducting research; 2) observer-as-participant, at which times I identified myself as a researcher and I did not participate in the groups' activities, and 3) strictly observer, at which times I neither participated in groups' activities nor identified myself as a researcher,[26] but merely observed events.

The following are singular events and categories of events which I observed in one of these three capacities:

1 Grassroots women's groups development and celebration activities—participant-as-observer;
2 Grassroots women's group meetings—observer-as-participant;
3 National NGO meetings on the NGO Coordination Act—observer-as-participant;
4 National MYWO meeting with Chinese women's delegation—observer-as-participant;
5 National Conference on Protection, Promotion and Support of Breast-feeding—observer-as-participant;
6 Christmas tree *harambee*—participant-as-observer;
7 Mothers' Hunger Strike—observer-as-participant;
8 Multiparty political meetings—strictly observer;
9 Ethnic clashes—strictly observer.

Procedures for data analysis

The procedures used for analyzing data emerged from both the normative school and the positivist school, though I relied more heavily on the former. The school referred to as the "post positivist" expressly explains the use of both qualitative and quantitative methods in this approach, albeit highlighting a heavier reliance on the qualitative. Postpositivism is fundamentally conventional.[27]

The approach that I utilized in this study was particularly useful for several reasons: 1) it underscores the importance for research to be carried out in its national environment, as naturalistic ontology stipulates that research problems are parts of wholes that cannot be understood properly outside of the context within which they exist, hence the necessity for field observation; 2) it is more useful for the kinds of data acquired in field research with human participants, as it acknowledges the ever-changing, non-predictable dimension of human behavior. In other words, it is more appropriate for qualitative research and open-ended data analysis; 3) it allows for the adaptability of the research design in the field as it realizes that every situational reality is different; 4) it is adaptable to the case study approach because it considers the particulars of a case important; 5) it sees the research problem as constantly being shaped by actions, reactions and interactions, such that there may not be isolatable specific cause(s) and corresponding isolatable specific effect(s); and 6) it acknowledges, unlike the positivist school, the constant role of values in conducting research—from choosing what problem to study, what paradigm and theory to use to guide the investigation, to what context within which to conduct the study. Overall, the post positivist approach is less constraining than the positivist approach, allowing for the holistic and human nature dimension of the problem to be considered.

The positivist approach also had much to offer. For instance, the positivist approach stipulates the following guiding principles, so that researchers are not merely asking questions which produce answers that become data which are not analyzable. The steps of the positivist approach which were most helpful in doing this were: 1) considering theoretical perspectives; 2) devising sampling and instrumentation; and 3) considering data analysis procedures. All of these steps were extremely important, though tentative.[28] They helped me to: 1) "make sense" of the data; 2) extract similarities and construct order from that data; and 3) identify themes and patterns which emerged from the data. Solid research that utilizes conventional qualitative methods intrinsically utilizes quantitative methods as well, albeit unappreciated by positivists.

The use of both the normative and positivist approaches in a post positivist vein was critical in enabling the research strategy to mature from hypothesis testing in its initial conceptualization to a more exploratory approach in its reconceptualization. Only a mix of the two approaches allowed me to capture the holistic problem as well as the cyclical nature of this research process in which recursive steps were repeated over time. During this process, I modified the data analysis procedures which I used and refined them to fit the research problem and the nature of the data I sought and collected. Moreover, I used both deductive and inductive logic, and I evaluated and re-evaluated theories and models of explanation. Overall, I used a multi-method research strategy.

PART IV: BENEFITS AND LIMITATIONS OF THE CASE STUDY AND MULTI-METHOD RESEARCH STRATEGY

Benefits

The case study approach to studying MYWO took an in-depth look at a national African women's organization, which at critical points in its history and interorganizational relations with its foreign donors and the Kenyan government claimed to be an NGO. This approach allowed an analysis of the shifts in the ways MYWO constructed its organizational identity vis-à-vis its development partners. For instance, MYWO has had a very fluid, chameleon identity, subjectively shifting between NGO and GO in attempts to maximize its patronage from foreign donors on one hand, and, at times, to maximize its clientage from the Kenyan state on the other hand. As discussed in Chapter 3, the MYWO elected leadership, during Wilkista Onsando's chairship in particular, hoped to maximize its clientage from KANU. MYWO under the leadership of Jane Kiano was less interested in clientage from KANU.

The case study approach further allowed MYWO to be observed and understood in the interorganizational relational context within which it operates daily. Observing both MYWO and its relational environment over time was important for dissecting development cooperation via partnerships, especially since MYWO does not exist and operate in a vacuum. Studies about development using the partnership approach are more realistic and offer more promise for examining development dilemmas today. Inevitably organizations, NGOs as well as GOs, which are forging linkages, collaborating and cooperating with each other, deal with contemporary global problems.

The case study approach of MYWO further allowed for the necessary focus on a consistent organizational phenomenon, allowing time for probing and sorting out inconsistencies in patterns that emerge in development cooperative endeavors. In addition, the case study focus on a women's organization allowed for the interrogation of gender relations as power relations (as stated in Chapter 2) involving "strategies and counter strategies of power" in the politics of development cooperation. The multi-method strategy made possible the probing into issues of MYWO's dependence, surrealistic autonomy, intra- and interorganizational linkages and its power vis-à-vis its partners—all in the wider context of development partnerships.

Limitations

There are also limitations to this study, some posed by political volatility, environmental uncertainty and the culture of fear and silence. Some are also reflections of the case study and the multi-method research strategy.

One of the limitations early in the fieldwork was resistance by MYWO headquarters to being studied. Some top national elected officials of MYWO were irritated by my presence and further irritated that I proposed to study MYWO as a resource-dependent organization in a development cooperative partnership. They were of little help in locating MYWO programs and projects and in making contacts in the field. For practical and methodological reasons, however, I interviewed the groups described in Part II of this chapter. These groups are not a random sample, nor are they necessarily representative of MYWO organizational linkages. Instead, they emerged from a mix of 1) the snowball method of interviewee selection, and 2) the research areas approved by the Kenyan government's Office of the President. In order to have a sufficient amount of data in general, and in order to have comparative data, I sought as interviewees a diversity of women's groups in the various MYWO programs, along with as many MYWO elected officials and staff persons as possible, at various levels within the organization and in as many areas of the country as possible. I also attempted to interview representatives of the entire population of foreign donors, since foreign donor organizations to MYWO were not many (see Appendix C). I secured information about locations of projects and about persons involved in MYWO's development interorganizational relations from various sources, which include newspapers, government publications, foreign donor publications,

Kiano and Shitakha administrations
tion), and from women with whom I
whom I was introduced during the
ees also provided information about
as well as other areas.[29]
olved the interviews—specifically the
sponses. Given the repressive nature of
d the culture of fear that had gripped
of my fieldwork, there were questions
re sometimes viewed as "too political"
ous. Moreover, questions about MYWO
ease that surfaced because of these ques-
interviews, caused the interviewees to
of getting referrals (snowballed) to other
etimes tried to broach the questions from
y assessment of the openness of the inter-
certain questions. This will be discussed

Cultu... also affected interviewee perception of the questions and th... instances, questions about "power" and "feminism" caused tremendous confusion (sometimes anger) and muddled some interviews. I eventually tried to find other creative ways to ask the questions, and sometimes I did not address problematic questions at all.

A final limitation is the restricted generalizability of the research results of this study. The results are particular to Kenya and to MYWO, and not generalizable to other African countries, or to other national women's organizations, or to African organizations that claim to be NGOs. This case study does, however, contribute to generalizability in the long run, for it provides these research results, based on a development cooperative partnership in Kenya as a general source of insights into the politics of development cooperation, gender and the reasons why organizations may make claims to their partners that they are NGOs.

5

RESEARCH FINDINGS: A BARRAGE OF CONTRADICTIONS

This chapter will present the findings of the study from each interviewee group responses: 1) women's groups, 2) MYWO elected officials and staff, 3) foreign donors, 4) Kenyan government ministry bureaucrats and 5) other relevant persons. Their responses will be presented in summary form. There are no end notes to this chapter purposely, because of promises of anonymity and confidentiality which most, though not all, interviewees requested. Also, the universal "she" will be used to refer to all women's group members, MYWO staff, ministry bureaucrats and other relevant interviewees, regardless of the interviewee's gender. Besides the universal "she" encompassing both the literal "she" and "he," the use of this form attempts to further protect the anonymity of the interviewees.

INTERVIEW SUMMARIES

Women's groups

The following is the summary of interviews conducted with 24 women's groups in Kenya, 6 of which were coalition groups of almost all women's groups in a particular sublocation, location or division. Also included is a business run by a Kenyan man—the leader of the business—and his family, because MYWO claims their business as its project. MYWO makes this claim because of the organization's connection to the family's work.

I found early in the fieldwork that there are no such entities as "MYWO groups." This is of particular importance for unraveling the confusion which surrounds MYWO and its relational environment. MYWO propaganda refers to MYWO groups, but technically there are no MYWO registered "groups." What exist instead are groups comprised of individual women, some of whom may be members of MYWO. Women often hold concurrent memberships in grassroots groups in their home area and in MYWO. Moreover, MYWO national headquarters registers individual women, not groups, although its constitution makes provisions for the registration of groups. MYWO national headquarters does not refute its reputation of having group memberships when it actually does not have them, for this makes its network seem larger to those it courts for assistance. Some groups even refer to themselves as MYWO groups. Many groups which had come to be referred to as MYWO groups did not even have projects which fell under the rubric of MYWO's four national programs. Because the word *maendeleo* literally means "development" or "progress" the number of women's groups who said that they were MYWO groups might actually have been development (*maendeleo*) groups of women. That they were development groups with women's memberships did not necessarily make them organizational members of MYWO or MYWO groups.

Of the 24 groups interviewed, the majority of them had some or all of their members who were members of MYWO as well. The "group" comprised of the business man and his family, who made pottery, including the Maendeleo jiko (energy-saving stove), had no members in MYWO. This family enterprise is presented as an MYWO group because of the training in making the jiko that the family provides to MYWO leaders and representatives. MYWO at the national headquarters does not make it clear that this family is not a MYWO group.

It is of further importance to clarify the women's groups' ties to MYWO, since there is another group which is confused with MYWO. That group is the KANU Women's League. KANU Women's League was created in 1989 as part of a KANU Directorate of Youth and Women's Affairs. It is considered an integral part of the KANU party and is subject to party control and discipline. It handles women's issues in development, politics and economic advancement. This group is very similar to MYWO in its appearance, activities, functions and relationship with KANU. A significant number of women seem to hold concurrent memberships in both KANU Women's League and MYWO, even though one does not necessitate the other.

Of the 24 groups interviewed, 2 indicated that they were KANU Women's League groups, and 3 other groups reported that they had members in KANU Women's League and in its leadership. It appeared that there was no blatant competition between KANU Women's League and MYWO in women's groups at the district and subdistrict levels outside of Nairobi. At these levels, outside of Nairobi, KANU Women's League and MYWO local leaders seemed to work together, particularly on issues of community development. MYWO local leaders appeared to make attempts to include KANU Women's League in its activities. KANU Women's League, like MYWO, seemed very intent on attracting foreign donors. The group that granted me my first women's group interview was a KANU Women's League group.

There was, however, tension and competition between MYWO national headquarters and KANU Women's League in Nairobi for favor with KANU and Moi. MYWO's chair, Mrs Onsando, had as her biggest competition Mrs Martha Othiambo, who was the "Mama" of KANU Women's League and a

MYWO provincial representative for Nairobi, and who was also closely linked to Moi. Moi and KANU exploited the competition between these groups.

The women's groups interviewed appeared to be relatively new groups. Of those who could recall their starting date, some reported having been organized in the 1970s, some in the 1980s and some in the 1990s. The groups seemed to have membership open to all women, although some required that eligibility be contingent upon having borne children. However, some groups had male members and men on their project staffs. There was also a great deal of flexibility in membership qualifications. Qualifications seemed to rest on participation in group activities and periodic payment of dues, both of which were not stringently imposed. As long as one paid what she could, she could be considered a member of a group. She also attended meetings when she could.

Membership in MYWO was more stringent, however. One had to pay annual dues to MYWO to be a card-carrying member of the organization. Moreover, to run for KMYWO office in the 1989 elections, one had to be a current card-carrying member of both KMYWO and KANU. For district and national offices, one had to be a life member of KMYWO.

With regard to the women's groups projects, an overwhelming number, 22, said that it was the group alone that formulated the ideas for their projects. Four of these groups stated, however, that they had consulted with various Kenyan government ministries, a Kenyan NGO (Christian Health Association of Kenya—CHAK) and a foreign donor (USAID) for assistance in the formulation of their ideas.

It was not surprising that the groups said that they had formulated their projects on their own when it was noted that *none* of the projects of the 24 groups fell directly under the four national program areas of MYWO—Nutrition, Leadership Development, Maternal Child Health/Family Planning, or Special Energy–Jiko. Although 10 of the groups' projects incorporated some of the elements of MYWO national programs, these had not been introduced to the groups by MYWO. In some instances, because the projects were implemented throughout the country, MYWO national office was not even aware of their existence. Elements of MYWO programs had been introduced either by Kenyan government ministry personnel or by foreign donors. Elements of programs similar to those of MYWO had also been gleaned from other groups with whom the women came into contact. For instance, one of the groups interviewed had copied the construction of a jiko for cooking from a DANIDA jiko project in their home area. In addition, several of the groups were assisted in their activities by a MYWO field worker employed by the Leadership Development program. In assisting the groups, the Leadership Development employee was implementing her women's training program. She was, not by design, introducing MYWO programs and projects to women's groups. She mainly assisted and advised whatever project(s) the groups were implementing. Perhaps because of the nature of the Leadership Development program, the only type of assistance MYWO could provide through its Leadership Development employees was indirectly related to the other three MYWO national programs.

The women's groups interviewed had a variety of activities ranging from gardening to teaching reading, from providing shelter for women and children to chipping ballast. Some groups had several projects functioning concurrently. Fewer groups had no functioning projects during the time of fieldwork, yet the women still met in regular group meetings. One of the common threads of all groups is that they all had income-generating components attached to their projects. They sold handicrafts, milk, water, fruits and vegetables and other goods. Although the groups attempted income generation, whether they made money or not did not seem to be their sole or primary determinant of success. They seemed to be more concerned with providing basic services to their community, and coming together.

Most groups considered their projects both successful and unsuccessful. Some of the common indicators of success they used were: 1) the benefits of the project to women; 2) reciprocal help between women; 3) the provision of goods and services to the community, e.g. bread, vegetables, education, health care; 4) good group participation; 5) sales. Some of the common indicators for lack of success were: 1) no market for their goods; 2) no sales; 3) loss of land; 4) financial and legal problems; 5) being forgotten by donors. Only two groups said outright that their projects were indeed successes. No group used the word "failure."

The women's groups' relationship to MYWO national headquarters and foreign donors seemed to be less institutionalized than MYWO claims, as well as less entrenched than the research initially assumed. The women's groups had no direct linkages to MYWO national office. For example, only 9 groups of the 24 groups reported having requested assistance directly from MYWO through its national headquarters or through their local MYWO elected representatives. None of these 9 groups received direct assistance. Four of these groups received training and motivational assistance from employees of the Leadership Development program, and one group received assistance from the MYWO national assistant treasurer who lobbied and networked on their behalf. MYWO national headquarters did not provide this assistance. It seemed that assistance to these groups was provided through Leadership Development individual employees because of the groups' ties to employees or MYWO leaders. Assistance was not provided because of specific objectives of the Leadership Development program. The national assistant treasurer complained that the foreign and Kenyan government assistance to MYWO and to women in general was being hoarded in Nairobi by MYWO elected officials. She searched for means to make this assistance more available and more widespread to other areas of Kenya as well.

Five groups reported not having requested assistance from MYWO: two groups because their strategy was to apply to the Rural Development Fund at the district level and gain MYWO leverage there; one group because their local elected representative had yet to visit their group and they were waiting to go through her, as protocol required; one group because they did not know how to contact MYWO; and one group because it was not a women's group although it had MYWO connections (previously mentioned family). Five groups could not recall whether or not they had requested assistance from MYWO, because their record-keeping had been neglected. Four groups were ambiguous or contradictory in their answer, so that I could not determine whether or not they had requested assistance from MYWO. There was one group of whom I could not ask the questions because of language barriers.

The women's groups seemed to receive more assistance from donors—either Kenyan government or foreign donors or both—than from MYWO at the national headquarters. A total of 13 groups reported having received some type of donor assistance. It is important to note two things here. One is that donor assistance to these 24 groups was not channeled through MYWO, even though Moi had declared in 1989 that MYWO would function as the focal and coordinating body for all assistance to women's groups in the country. The other important thing to note is that the type and amount of assistance these groups really received from donors must be qualified. Assistance included anything from grants to loans to bicycles to skill transfer seminars to cow sheds. Grants and loans also varied widely in range—from Ksh 3,000 to Ksh 150,000 (equivalent to US $100 to US $5,000 based on early 1992 exchange rates)—according to what was reported.

It appears that groups may have received either "ongoing" assistance or "push/crisis" assistance or both, from foreign donors or Kenyan government Ministries. "Ongoing" assistance I defined as continual assistance for the specified duration of a project. "Push/crisis" assistance I defined as usually one-time intervention assistance, either to begin a project or to help at a critical point for the survival of a project. Nine of the groups reported having received some type of ongoing assistance—financial, technical, social or material—from either foreign or Kenyan government donors, or both. Only 3 groups seemed to have a schedule of assistance and mandated field evaluations by foreign donors. These groups also seemed to get more financial assistance. Nine groups reported having received, at some point, push/crisis assistance. Within this group of 9, there is substantial overlap with the aforementioned group receiving ongoing assistance. The groups that did not receive donor assistance survived on members' contributions.

There were three questions which posed particularly difficult problems for collecting data. Two sought similar data regarding Kenya's leaders in development. The questions are: Which of the partners took the lead in development? Was there a hierarchy or equality between partners? Who should be credited for the overall development of Kenya?

With regard to the first question, most of the groups did not answer. Of the ones who did, all of them cited the government of Kenya and various ministries and foreign donors as taking the lead in Kenya's development. Some groups added women. Only one group ranked a hierarchy—the Ministry for Culture and Social Services (MCSS) at the top, followed by MYWO and then the government respectively. MYWO was mentioned only in this response. The same answer—that the government of Kenya and various ministries and foreign donors took the lead in Kenya's development—was reiterated by most groups for the second question, with the addition of the government's National Development Plans by one group.

It was no surprise that these questions were problematic given the political climate within which this research was taking place. As Chapters 3 and 4 discussed, the government and donors were at odds on issues of funding, accountability, responsibility, transparency, human rights and, in some cases, sovereignty. In more sensitive environments within the country, and at times when donor–government relations were particularly precarious, the questions may have been reworded in attempts to depoliticize their nature and still solicit the information.

The third question which posed some problems was whether or not the women's participation in the project had increased their power as women. For several reasons this question may have been problematic: 1) the concept of "power" seemed to have a different, perhaps cultural, connotation; 2) there was a reluctance to answer the question in front of men who may have been present; and 3) often women did not want to be explicitly associated with politics. When I mentioned "power" the women seemed to automatically relate it to state politics in Kenya.

Most of the groups who answered the question did not use the word "power." They instead said that they had increased "rewards" or "satisfaction" from participating in projects, such as more food, more educational security for their children, more money and more bargaining strength with their husbands. One group, from North Eastern Province, said that they had always been "stronger" than men in their area and "stronger" than women in other areas of Kenya. Only three groups used the word "power." They said that they felt an increase in personal as well as public power, for women were now running for political office and serving as judges. One group said that they felt no increase in power because they had always been powerful.

A comprehensive view of the interviews with the women's groups adds a dimension not considered by interorganizational relations. That is, there were very tenuous linkages, when there were linkages at all, between the local women's groups and the focal organization, MYWO national office.

The linkages seemed much stronger between the local women's groups and the ministries, and the local women's groups and some foreign donors, particularly NORAD who ceased being an official donor to MYWO in 1990 after the Wamwere incident discussed in Chapter 3.

For this group of interviewees it must be noted that particular problems may have affected the responses and their interpretations. There were language barriers with two groups, which affected parts of the interviews. One group's project was in the middle of an ethnic land clash during my visit there. Moreover, with two other groups, the MYWO district chair was present during the interviews and responded often on the group's behalf. With one group, the district social women's officer (DSWO) of the Ministry for Culture and Social Services was present and challenged the answers of the group in my presence, especially their praises of Jane Kiano. In other instances as well, other persons including politicians, government officials—men—and women leaders may have been present during the interviews, and their presence may have affected interviewees' responses and my interpretations.

MYWO elected officials and staff

Elected officials

The following is a summary of interviews with 10 MYWO elected officials. All the interviewees agreed that the general role of MYWO was to assist women throughout the country, especially at the grassroots. Only one interviewee responded that the role of the organization was to meet the national development needs of Kenya. There were varying views about the ways in which women were being and should be assisted, demonstrating that MYWO elected officials did not have a monolithic purpose. As one interviewee stated, "MYWO's purpose depends on who the leader is." Two of the current elected officers said that it was the organization's purpose to improve the economic standing of women, for it was the belief of many, though not all of the Onsando administration, that "money is the key to empowerment." From the interviews, it seemed that the diversity and lack of clarity of purpose of the organization may have been problematic for the functioning of MYWO.

From the interviews it seemed that MYWO had shifted in focus over its life-span as a women's organization. One interviewee outlined the following changes in MYWO's role over time as:

1 the *civilization* of African women by Europeans;
2 the raising of African women's standards to that of Europeans;
3 catering for African women through development and income-generating activities.

This interviewee identified phases of MYWO's evolution up to Kiano's resignation in 1984. She and another interviewee agreed that, following independence and during the Kiano administration (1971–84), MYWO functioned as a welfare and development organization whose focus was nation-building and development in the spirit of *harambee* ("Let's all pull together"). Today, they said, by contrast MYWO's focus has changed. It is not nation-building.

These interviews with two veteran members of MYWO from the Kiano administration demonstrated that MYWO's national headquarters linkages to women groups, especially rural women, were also different in the 1990s. These interviewees recounted how MYWO used to provide stoves to needy families and wheelchairs to the disabled. They stated that, in the past, persons assisted by MYWO did not necessarily have to be members of the organization. Today, however, according to the current MYWO president (Onsando), local women must be affiliated with the national office of MYWO in order to be assisted. That is, local women must have purchased a current membership with the organization or be affiliated in some other way to be assisted. One other way to be assisted, as I observed, was to live in an area where there was an active employee of the Leadership Development program who was assisting and trying to revive women's programs. One of the interviewees from the current administration stated that MYWO's revival efforts in her areas were fruitless because the educated and wealthy women were not interested in joining MYWO, and the poor were too poor to buy membership cards to join.

One of MYWO's national headquarters' main selling points for securing foreign donor funding is its claim to work to assist economically marginal women. Ironically, a membership fee is required from these women to qualify for assistance. Hence, a major question arises from this and was asked of the interviewees: Do women representatives of MYWO know where the foreign donor funding goes when MYWO national headquarters secures funding to assist women? All except one of the interviewees at the provincial, district and divisional level said that they did not know anything about MYWO's dealings with foreign donors. Many of them did not know the names of the donors who funded MYWO's current national programs. Their exposure to donors and to matters between foreign donors and MYWO was very limited. Interviews indicated that the women seemed uninformed of their rights in MYWO, or too timid to express them. Some of the interviewees further added that MYWO raises the economic hopes of women by its propaganda campaigns, but does not meet them.

The national assistant treasurer of MYWO, in our interviews, did raise questions about the allocation of foreign donor funding as it was not trickling down to her area. She also raised questions as to the whereabouts of the Ksh 50,000 allocated to women's groups in each district in Kenya by Moi in 1992. Surprisingly, however, as interviews and discussions indicated, in spite of her being the national assistant treasurer she had no knowledge of the monies of the organization and was generally not privy to its financial affairs. It seemed that her election to the position of national assistant treasurer was more symbolic than substantive. The only persons who appeared to have knowledge and decision-making power with regard to the finances in MYWO were the president, the CEO, the secretary and, to some extent, the national accountant.

During our interview, the MYWO chair would not speak of the finances of the organization or decision-making power with regard to the finances, despite probing questions. Instead, she talked of other organizational matters, albeit very superficially. She was uncomfortable, restless, over-busy with other things simultaneously and would not focus on our interview. Some of the major complaints which she, as well as other interviewees, raised with regard to MYWO's relations with foreign donors were that:

1 Donors place too many restrictions on foreign financial assistance;
2 There is too much imposition of foreign donor ideas, not allowing MYWO enough say;
3 Too much of the money goes to foreign donor preferred areas; and,
4 Too much of the money goes to MYWO national program officers' preferred areas.

She lacked examples to support many of these complaints, especially the first two. Interviewees who made these complaints, including the chair, could not really provide concrete instances, or documents as evidence to support the charges, although some of the allegations of donor and ethnic preferences for certain areas seemed to have some degree of legitimacy.

From the recollections and comparisons by interviewees, the Onsando experience with foreign donors seemed very different from the Kiano experience. Whereas under Onsando, relations seemed to be strained with many donors, interviews from veteran MYWO members stated that, in the 1970s, MYWO under Kiano had easier cooperative relations with donors. They stated that there had been no pressure or monitoring from donors, and that MYWO enjoyed a great deal of flexibility in utilizing the assistance it received. They stated that foreign donors took its direction from MYWO, and MYWO implicitly took their direction from the Kenyan government. One interviewee put it this way, "Most of the work of NGOs (Kenyan and foreign) was complementary to the work of government, although there was not formal cooperation."

With regard to MYWO relations with the Kenyan government, most of the interviewees stated that there was "good" cooperation with government, while they clearly made a distinction between the ministries and KANU. The ministries' work was the work that was deemed "good." However, KANU—the party—according to many of the interviewees had caused some problems. For example, some interviewees blamed KANU for the loss of foreign donors, and some felt that KANU had not been helpful to MYWO in reaching its aims at all. One interviewee was very staunch in disagreeing with the belief that KANU had caused problems for women. She stated that KANU had been most helpful in the 1989 elections in "protecting women from foreign ideas." She was closely linked to KANU. Some interviewees agreed with her that KANU had not caused problems, but they were not as staunch in their response.

Not many of the interviewees wanted to address the issues of development cooperative partnerships or state whether or not a hierarchy of partners or equality between partners existed in the partnerships. At the lower administrative levels—division, district, province—the interviewees seemed to have limited awareness of partnerships. Thus, most did not discuss partnerships because of this. Only 3 interviewees from the current administration answered, all with differing perspectives. Two of the interviewees said that a hierarchy existed. According to one, women were at the apex (contributing at least 80 percent of national development), followed by government, then foreign donors. According to another, government was at the apex, followed by many institutions including but not exclusive to MYWO and foreign donors. She did not rank these. The chair of MYWO was very vague in her answer. She said that the partner who leads depends on what is needed and this determines what the partners give. She added that when partners cannot agree on program formulation and implementation, generally it is MYWO who is forced to bend to the pressure of foreign donors. Another interviewee voiced a similar view, and stipulated that when MYWO does not bend, it is depicted unfairly as a "culprit."

The interviewees from the Kiano administration said that, in the 1970s, the government was at the top of the hierarchy of all development partnerships, because nation-building needs were the priority of all. They agreed with each other that the nature of development relationships changed with the presidents of the country, and perhaps this partly explains why MYWO is different today compared to Kiano's time. One of the interviewees made the following comparison, to paraphrase: During the tenure of President Kenyatta (1963–78), the focus of the country was development. With President Moi (1978–present), this focus had changed from cooperation to politics and control. These two interviewees felt that development partnerships had become too politicized, mainly because the government wanted control and was greedy.

Most of the current elected officials agreed that it is the national office of MYWO and local groups that formulate the ideas for the MYWO programs/projects that are implemented. The chair did not acknowledge the role of MYWO's national office in program and project formulation at all. Instead she identified local women, district commissioners (DCs), and district development committees (DDCs) as the formulators of MYWO programs and projects. She also said that local women may propose ideas which the national office merely writes into a proposal.

Most elected officials identified no tangible and measurable indicators for project success or failure. Moreover, they were very vague as to whether or not programs/projects they had implemented had been success or failures. Of the current administration, only the national chair claimed overwhelming success for projects. Her indicator was the vast membership of MYWO, a reported 1.5 million, for which there are no current statistics or records. Of the elected officials who did discuss indicators of program and project success or failure they seemed to look at indicators other than whether or not groups had reached their specific objectives. Some indicators of success which they voiced in interviews included: 1) having something to do other than kitchen work, and 2) the continued motivation of the women to remain in groups. From the interviews it seemed that many women felt that if their projects did not survive or were not successful, it was not their fault. They blamed lack of program and project success on not having resources. It is important to note that many of these women, especially at the national level, were not involved in any hands-on development projects. Many were politicians, who talk about projects, but who were not involved in hands-on project/program implementation.

The interviewees from the Kiano administration claimed outright success for their projects, unlike interviewees from the current administration. One of the interviewees from the Kiano administration said that a major reason for success of MYWO under Kiano was that MYWO was "not racial or tribal." The other interviewee from the Kiano administration said that one of the things that led to her disillusionment with MYWO and the failure of MYWO programs/projects was that most of foreign donor funding to MYWO was being allocated to salaries and transport. Very little money was actually going to the projects. For these reasons, she decided to leave MYWO.

Instead of focusing on evaluating the success or failure of programs/projects, the current MYWO organization seemed to be riddled with problems. Many of the interviewees discussed or alluded to these problems. The chair either denied or did not acknowledge them. From the interviews, it seemed that the problems which led to the current crisis of the organization were: 1) the financial mismanagement by Mrs Shitakha, who was the chair of MYWO during the mid-1980s; 2) the affiliation of MYWO to KANU in 1987—an affiliation which Kiano had reportedly, according to one interviewee, resisted for 12 years; 3) the rigged elections of 1989 during which time KANU women who were staunch Moi supporters were placed in office; and 4) donor withdrawals from MYWO in the 1990s. One interviewee indicated that the organization had become too large and its enormous size had caused these problems. Superimposed on all of the problems seemed to be MYWO's current national leadership's hunger for political and economic power, with the exception of the national assistant treasurer. Not all of the interviewees stated all of these reasons, but each interviewee stipulated at least one of the above reasons as a cause for the organization's crises.

Most of the interviewees would not address the affiliation of MYWO to KANU, and its subsequent disaffiliation. Of the interviewees who did respond, some felt that the affiliation was political maneuvering by KANU to control women, which also caused unanticipated donor withdrawals. Most interviewees welcomed the disaffiliation. The chair and another district official disagreed with the disaffiliation. The chair did not regard affiliation to KANU as a problem for the operation of the organization. The district official who felt that KANU protected women from "foreign ideas" also thought that the affiliation was a positive move, for, according to her, "it is not possible for women to function independently." She further said that MYWO was not, should not and could not be an autonomous body.

This district official's view of not wanting autonomy for MYWO "because women should not act alone" did not correspond with the other interviewees who responded to the question. The chair claimed that MYWO was, in fact, autonomous, even during its affiliation with KANU. She said that since her election in 1989, "affiliation, disaffiliation has not made much of a difference . . . MYWO has not operated with KANU despite the name KMYWO." Another interviewee's response diametrically opposed this. She stated, "MYWO is KANU." For her, KANU was so subsuming that even her women friends were not allowed to be women disliked by KANU, because she was KANU. In her interview, she spoke of Wangari Maathi, the world-renowned professor, activist and environmentalist, whose work the interviewee respects and admires. She lamented on how her area could have benefited from Maathai's Green Belt Movement in its fight against erosion. She could not, however, be friends or have a working relationship with Maathai because of the different nature of their relationships to KANU. Whereas the interviewee is a member of KANU, Maathi is KANU opposition and takes KANU to task for its ills as described in Chapter 3. Regarding the autonomy of MYWO, the interviewee did not think that MYWO was not autonomous because "it could not be." Instead, she said, MYWO was not autonomous simply because it was a part of KANU. Although she was one of few women KANU members, she welcomed MYWO's disaffiliation.

This interviewee felt that MYWO had assisted in increasing women's personal power, as well as women's confidence to run for public office. She added that men were not happy with successful women's projects. Only one other interviewee agreed with her and used the word "power" as well. The other interviewees talked about social and economic improvements women had experienced—which they attributed partly to MYWO.

There were several recurrent recommendations which emerged from the interviews for more effective approaches to cooperative development partnerships. They include: 1) more dialog between partners; 2) identification of *real* Kenyan NGOs; 3) publication of donor financial records, for they had no idea how much assistance MYWO received from its donors; 4) allowance of foreign donor access to grassroots without going through government or MYWO national headquarters; and 5) keeping men out of women's affairs. The second recommendation is very interesting, considering MYWO has dubious status as an NGO. One interviewee even called for new and legitimate elections for MYWO.

From these interviews with elected officials, several issues became apparent: 1) the majority of the interviewees were not concerned with creating self-sufficiency for the organization; 2) there was a great dependency on foreign donors, not only financial and technical dependency but also, and maybe to an even greater extent, psychological dependency; and 3) to some degree, securing a foreign donor was considered a success in itself. Only one interviewee talked about creating wealth for the organization from within, to the extent that she would not spend time answering questions that dealt with donors—foreign agencies or the Kenyan government. The interviewees did not recognize the Kenyan government ministries as donors.

From the interviews, it became obvious that MYWO national office seemed to have few and weak network linkages with local elected officials. Many times, elected officials from the lower-level administrative units did not seem to be aware of what the national MYWO was doing. Many times, MYWO national level did not seem to care what lower-level administrative units were doing. There seemed to be little unity and coherence in MYWO's intraorganizational inter-level leadership. Moreover, the national chair seemed to be detached from Kenyan women.

Throughout these interviews, two of the interviewees were quite restrained in their answers, as their district chairs often assumed the prerogative of answering on their behalf.

Staff

The following is a summary of interviews with 16 MYWO staff persons. In this set of interviews, there were 5 male interviewees. Including men in MYWO staff interviews was not expected, but men occupied staff positions and agreed to interviews. It was noted that men were usually hired to perform the technical work, i.e. accounting, budgeting and evaluation of the organization.

When asked of MYWO's role in development, most of the interviewees said that it was to assist women in Kenya. Two national program managers gave answers more specific to their programs, i.e. Maternal Child Health and Family Planning (MCH/FP) and Leadership Development (LD). The MCH/FP program manager said that MYWO's role was to provide family planning so as to assist the government's efforts in reducing population growth and providing good health for mothers and children. The LD program manager said that MYWO's role was to develop leadership of women, improve living standards for women and voice the ills of women.

Most interviewees referred to their job descriptions, albeit unwritten, to describe their specific roles as staff persons in MYWO. For instance,

1 The SEP–Jiko program manager stated that it was her role to disseminate *jikos* and follow government–Ministry of Energy guidelines for the implementation of this project;
2 The LD program manager said that it was her job to give training, motivation and skills, and to build the esteem of women so that they could fight for their rights and withstand being patronized by male government leaders;
3 The MCH/FP program manager said that it was her role to implement the program according to the government's national development plans, and to try to fill the gaps for family planning and child health that the government could not reach.

Two of the program managers' roles were to implement programs according to the agendas of the Kenyan government—SEP–Jiko and MCH/FP. The LD program officer was more activist-oriented in her role. The Nutrition program manager did not have a current role in the program, because the program had ended. She was still on MYWO's staff.

Most of the staff persons interviewed seemed to be the actual linkages in MYWO's relational environment to the women at the grassroots. Many of them identified target areas and traveled, or delegated representatives to travel, to project/program sites. Each program manager and her staff met women at the different administrative levels—province, district, division—though irregularly. Other staff at the headquarters, such as the archivist and the accountants, were to keep records of MYWO activities, including activities at the grassroots.

Only two of the interviewees stated that their programs were formulated by MYWO, one of which was created by Mrs Kiano (LD) and the other (Nutrition), created by the program staff. The latter project had ended at the time of my fieldwork. The other two national programs—MCH/FP and SEP–Jiko—were created by the Kenyan government; for both of these the program managers either did not know or did not give specifics about their formulations. The other staff interviewed said that they had no information about who was involved in the formulation of the programs. They merely came to work and did their jobs. Most of these interviewees seemed to be technocrats and professionals who did not ask questions about MYWO. They merely performed their duties.

Little was known among the staff about program and project changes. Most interviewees deferred to the national program managers when asked about changes. National program managers were unclear about how changes were dealt with in the Onsando administration, other than the fact that meetings were sometimes held with foreign donors "to sort things out." They said the elected officials often did not abide by the decisions made at these meetings. The program managers blamed the MYWO elected officials for the uncertainty about organizational matters, arguing that they were non-communicative, inconsistent, uninterested in the organization and aloof as to how interorganizational relations should be handled between MYWO and its donors. The national program managers were fully aware of the threats MYWO elected officials posed to the development partnerships of MYWO.

Only a few of the MYWO staff outside of the headquarters at the Nairobi office seemed to have a comprehensive view of the actors and interactions in MYWO's relational environment. Lower-level staff in and outside of Nairobi may have vaguely known of some of the partners in development, if the donors visited their particular site. They seemed to know less about MYWO's national office interactions with its foreign donor partners. The sub-national staff interviewees were more concerned with the day-to-day organizational activities in their particular area. Their interaction with MYWO national office with regard to their particular program seemed to be mostly top-down, on occasions when there was communication. The local staff, from the grassroots, were less likely to approach MYWO national office. They seemed to wait for MYWO national office to come to them. There seemed to be deference and timidity on the part of most of the women at the grassroots, and arrogance at the national level within the elected leadership. There was also a great deal of friction and resentment between the national program managers and the national elected officials. To the national program managers, these elected officers "slowed down progress" and were a hindrance to development, particularly since program managers were required to have all of their program's activities approved by elected officers and the CEO. One program manager remarked that MYWO's development programs had been held back since the 1989 elections because of the current elected officials' incompetence and lack of interest in women.

Most of the program managers felt that the government was at the head of development partnerships. To some, this was positive, to others it was not. One program manager felt that the government was at the top of the hierarchy and that the partners in the program worked well together in partnership. Another program manager felt that the government was at the top of the hierarchy, but that something other than partnerships existed. She felt that the relationship between foreign donors and MYWO was strong in terms of implementation of the program, and that government intervened only on policy concerns. A third program manager agreed that government was at the top of the hierarchy, but that instead of a cooperative arrangement existing between partners, there existed opposition. She stated that the government was against both MYWO (excluding the elected officials) and foreign donors because it wanted to remain corrupt. She argued that the KANU government did not want to share power, therefore it was against multipartyism and all of its suspected proponents.

A staff person further revealed very important information on foreign donor support to MYWO. She stated that 79 percent of all of MYWO's operations are funded through Pathfinder International, who receives 93 percent of its funding from USAID, thus placing USAID at the top of the hierarchy of the partners in development in terms of resource contributions to the partnership and organization. Because of this, she said, USAID called the shots for MYWO. She also considered USAID her employer, not MYWO. She said that she was inclined to follow USAID's orders to the letter because she did not want to face unemployment. This shed some light as to why USAID yielded considerable power with MYWO and Pathfinder (to the extent that they could summon personnel to meetings at the drop of a hat, even if it meant interrupting other meetings) and with other US foreign donors engaged in family planning programs/projects as well. For example, all US donors to family planning projects must get clearance from USAID. This is reportedly US policy relating to development funding in Kenya. Moreover, the MCH/FP program manager discussed the restrictions placed by USAID on development finances because of US laws. For example, assistance to the MCH/FP program cannot be used to fund abortions. The program manager felt that this was a case of US law being imposed on Kenya, and claimed that foreign donors, particularly USAID, wanted to choose sites for program and project implementation where they could get quick results and successes. This often caused overlap with NCPD and project duplication for the same target areas and groups.

There were conflicting views among the staff about the role the government has played in development partnerships. On one hand, the national program managers who implemented programs initiated by the government felt that the government had a development master plan, which was to be followed by MYWO and foreign donors so that the country would develop. On the other hand, some staff interviewees, including one program manager, felt that government's only plan was to maintain political, economic and patriarchal hegemony at all costs. These interviewees asserted that there was too much control and corruption by government. They felt that KANU's affiliation of MYWO in 1987 was to take power from MYWO and cause confusion among women. This confusion, they argued, facilitated the exploitation of the MYWO elected officials by KANU. One of the interviewees went further to say that *all* of Kenya's problems, not only MYWO's, were created by a corrupt KANU and Moi. The interviewees saw the ministries as separate from "government."

KANU was definitely blamed for the withdrawal of donors from the organization and for the lack of women's interest in joining the organization.

KANU was also blamed for the unemployment of over forty former MYWO staff who were displaced when NORAD was kicked out of Kenya by President Moi. Moreover, KANU was accused of hijacking MYWO by the affiliation in 1987 and by rigging their elections in 1989. Some interviewees either stated outright or implied that KANU had placed hand-picked KANU women in office to secure donor funding for it and to serve as its mouthpiece to women and to foreign donors in support of government policy. Though not all interviewees would speak on this subject, some believed that KANU was exploiting women for its own ends and that the elected leadership of MYWO was allowing a women's institution to be exploited for their personal gains of money, power and status within the male-identified and dominated KANU party.

One of the program managers welcomed the disaffiliation, as she felt that MYWO should never have been affiliated with KANU in the first place. She hoped that disaffiliation might bring an end to the confusion between elected officials, program managers and grassroots women. Moreover, several of the interviewees hoped that disaffiliation might bring donors back to the organization. One interviewee observed that "donors are confused as to the autonomy of MYWO. They do not understand MYWO's position." As relayed by two of the program managers, many donors refused to support a political system, therefore the donors left MYWO and Kenya. For example, one program manager said that US donors have stated that their money will not support political activities. (Perhaps this is one of the reasons that any funds directed to the MCH/FP program from a US-based NGO must be cleared through USAID.) Ironically, USAID, via Pathfinder and other US foreign donors, has continued to increase its financial assistance to MYWO, in spite of MYWO's affiliation to KANU.

Most of the interviewees agreed that MYWO was not autonomous. Two of its national programs are directed and controlled by government, while MYWO is part and parcel of KANU. Only one of the interviewees argued that although MYWO was affiliated to KANU it was not a part of KANU. She stated, "The chair [of MYWO] sits with KANU only to represent the interests of women."

The MCH/FP and LD program managers felt that their programs were successful. The MCH/FP program manager enumerated some obstacles which stood in the way of increased success: money, resources, donor withdrawals and demand for services by women being greater than MYWO's ability to supply. The LD program manager believed that women had increased self-esteem and power as women. The SEP–Jiko program manager did not know if her program was successful or not, because MYWO did not conduct evaluations. She was satisfied with this.

Three of the field workers and one accountant felt that the projects in their areas showed some signs of success. Among their indicators for success were: 1) the revival of projects; 2) women showing more confidence;

3) women in greater control of how many children they had, and subsequently their ability to educate them; and 4) the increasing numbers of women who wanted family planning services. Some interviewees, particularly one district accountant, felt that tensions spawned by multiparty politics had caused a problem for the successful implementation of projects and the delivery of supplies to certain areas. She expressed concern about the long-term effects of multiparty politics on development. Others welcomed multiparty politics. They thought that if there were multiparty politics, MYWO might regain its pre-affiliation status.

Some interviewees felt that women had increased their power because of their involvement with MYWO. They said that women had more control over their bodies and childbearing. They felt that the MCH/FP program gave them a better chance to make a living. Most did not use the word "power", however. There was only one interviewee who used this word—the LD program officer. With regard to the women in the LD program, she said that they had increased their power by learning their rights, hence making it more difficult to be used by men. One program manager felt that women should be skeptical of the NGO Coordination Act. She felt that it was another instrument to control society, and women in particular, contrived by the government.

There were many suggestions for more effective development partnerships. Some of the suggestions are for particular programs, fewer are for MYWO in general. One program manager felt that the foreign donors should bring the most lacking of all necessary resources—water—because its shortage was a major technical resource problem. Another program manager felt that foreign donors make too many demands for little money, particularly the foreign donor CEDPA. She felt that there should be fewer demands and more financial support for MYWO. In addition, she insisted that foreign donors should establish offices in Kenya, as opposed to remaining overseas, in order to facilitate communication, coordination and cooperation. Moreover, she stated that money needed to arrive on time. A third program manager said that the Kenyan government needed to leave MYWO alone, and that government should not bully foreign donors. She asserted that foreign donors should put more pressure on the Kenyan government to end repression and human rights violations, and that Kenyan citizens should take the fate of their country and their future into their hands. To paraphrase, she stated, "if donors leave and the government falls, it falls, and this Kenya has to deal with." She was frustrated that women in particular did not have an open environment within which to work. She stated, "When you start talking politics, they [men] think you are after their blood." She also added that women at the grassroots need expertise, not money, and donors should be aware of that. Her assistant added, and she concurred, that donors should not be averse to paying salaries to the implementors or staff of the program.

In the LD program, the program manager, assistant manager, program secretary, program driver and 22 field coordinators were receiving regular salaries, 25 percent annual gratuity and leave allowance. All were enrolled in a National Social Security Fund (NSSF) and comprehensive group insurance. The program manager and assistant manager were receiving a housing allowance. This was paid by KAF. In the MCH/FP program, the program manager, assistant program manager, program secretary, program driver, an unknown number of district coordinators, divisional coordinators, accounts assistants, copy typists, watchmen, nurse coordinators, office messengers and approximately 500 community-based distributors (CBDs) were paid regular salaries and received annual gratuity and leave allowance. They were also enrolled in NSSF and group insurance funded by the MCH/FP foreign donors, of which USAID was the major donor.

A few caveats are necessary here. My interviews with the Nutrition program manager really did not focus on this study. She was more interested in discussing gender relations among African Americans, such as the Anita Hill–Clarence Thomas case. I did manage to salvage some information. As I delved deeper into my research, I discovered that the Nutrition program manager was in quite an unstable position with the organization. Her program had ended, and the CEO and chair were trying to oust her. However, as a 13-year veteran of MYWO, she was successful at being requested by the Population Crisis Committee of Washington DC to participate as an advisor and field worker in a study on female circumcision called "Harmful traditional practices that affect the health of women and children" which was being conducted through MYWO networks in 1992. Perhaps her uncertain future with the organization influenced why we never talked at length about MYWO. After she became involved with the female circumcision study, our paths did not cross again, although I attempted several meetings with her.

In two interviews with assistant program managers, very scant information was obtained directly, since they usually referred me to their program managers, perhaps because their knowledge was limited or they feared saying something contradictory. In two other interviews with the Machakos District accountant and the Mombasa Leadership Development (LD) field worker, the interviews were not private and the interviewees really did not answer very many questions, either because they appeared timid or because some "authority" answered on their behalf or diverted them from the question at hand. In some instances, they simply did not want to talk about MYWO while they were conversant about other things. Of the MYWO staff persons, two key people in MYWO refused to talk with me—an appointed official and a field worker for Muranga District. They were angry at my presence and despised the fact that I was doing the research on MYWO. Both gave me the run-around and a very hard time when I tried to talk with them. The field worker asked me to buy her a new set of tires for her MYWO vehicle so that she could consider whether I should be granted an interview.

Foreign donors

The following is a summary of the interviews conducted with foreign donors providing assistance to MYWO. These donors—USAID, the World Bank, US Peace Corps, Pathfinder International, CEDPA, GTZ, KAF, Marttaliitto and Coca-Cola—are referred to in Chapter 4 and described in greater detail in Appendix C. These donors provided one or more of the following to MYWO projects/programs: financial assistance, technical assistance and training, material assistance (i.e. supplies, materials) and educational assistance and training. Most of the donors did not reveal the amount of financial assistance they provided to MYWO for the implementation of their project/program.

As foreign donors, their roles in Kenya were very broadly either humanitarian, environmental, diplomatic and/or political. That is, their projects were geared toward the promotion of human welfare, protecting the environment or nurturing so-called good relations between nations, as well as and sometimes concomitantly with promoting "democracy" and capitalism.

One of the reasons these donors gave for choosing to operate in Kenya is that Kenya has historically had a conducive environment for donor relations and activities. Donors could gain relatively trouble-free access to Kenyan NGOs and government. Moreover, these donors could justify the implementation of their projects in Kenya to their home governments and agencies. In addition, as they pointed out, expatriates were welcome and could function in Kenya with few problems. Kenya has, since independence and perhaps until recently, been considered a favorite African nation by Western donors.

MYWO was reportedly selected by the foreign donors in this study as the focal organization with which the donors would interact principally because it was believed to be an organization which had established networks throughout the country, and also because it had been delegated by Moi, in 1989, as the national center for all women's groups. The women's groups to which MYWO headquarters was reportedly linked had the reputation of being dynamic. They were believed to have the potential to interlink urban and rural, province and district, older and younger, and rich and poor, in the national development of Kenya. For many foreign donors, MYWO had already established the necessary infrastructure for the implementation of large-scale development programs brought to Kenya by foreign donors. As Marttaliitto stated, as a foreign donor its basis for choosing MYWO was that it decided to "leap over when the fence was lowest." Not all of the foreign donors were as blunt about their choice of MYWO as Marttaliitto.

The most common short-term overarching objective of the foreign donors as a group, seemed from their interviews to be the improvement of women's health and life chances through population control (family planning), health care access, economic and social empowerment, and environmental conservation. The long-term goal of these donors seemed to be the "development" of Kenya, and Kenyan citizens' acceptance of liberal democratic and capitalist ideals.

In order to reach these objectives and goals, the majority of donors appeared to have brought with them already formulated ideas, projects and programs geared toward making women's health and life chances better. Thus, these donors expected projects and programs to be implemented as they had been formulated in the foreign donors' home countries. KAF was the most obvious exception to this. It had no obvious agenda attached to its financial assistance. Marttaliitto, to a lesser extent, did not have a blatant foreign donor agenda either. With regard to the other foreign donors, however, MYWO seemed to be a conduit for the implementation of their already formulated projects and programs. Some donors also taught, in their seminars, the fundamentals of creating and maintaining "democratic" political structures and the financial benefits of entrepreneurship. Of the donors, it was the Peace Corps that focused most directly on creating and maintaining democratic political structures; and it was the Peace Corps and Coca-Cola that focused most on promoting entrepreneurship. To varying extents, all of the donors, with the exception of GTZ, focused on fostering entrepreneurship.

Overall, the donor institutions did not seem to have respect or particularly high regard for MYWO. They simply saw MYWO as a vessel for their project/program. As KAF stated, "Pathfinder merely uses MYWO as a cover to do good work. Pathfinder has the organizational capacity to run a project. They have management, administrative and recruitment teams, and adequate personnel." KAF added, "Pathfinder must operate in spite of MYWO . . . MYWO is a lot of propaganda. It is nothing." The World Bank interviewee made similar observations about USAID. She indicated that USAID had large in-country offices, and its staff played a direct role in the implementation of the programs and projects that the World Bank funds.

MYWO's reputation of having a functioning national network of women's groups (earned more unjustly than justly) was clearly its most important asset for attracting foreign donors. Donors seemed to be attracted to MYWO because of its past "reputation." MYWO of the 1980s and 1990s had attracted only three of its current donors—Coca-Cola, KAF and Marttaliitto—through the submission of proposals or formulations of projects. The idea for the Coca-Cola project was not new. It was merely reintroduced to Coca-Cola by the current chair, Mrs Onsando. A similar project had been implemented during the Kiano administration. The Leadership Development project with KAF was also introduced by Mrs Kiano. The proposal for the Nutrition project with Marttaliitto came from the MYWO national office, in collaboration with Marttaliitto. All of the other donors brought their agendas with them from their home offices, already formulated. Peace Corps volunteers may be an exception because they do some limited negotiations with their clients. They, however, have an overall job description and organizational goals that they must fulfill as volunteers. These findings suggest that MYWO's program agenda was not indigenous, for the most part. Certainly, it was not grassroots.

Furthermore, there seemed to be few agenda changes in MYWO's projects and programs. Negotiations between MYWO and donors before project implementation was initiated were reportedly almost nonexistent. It seemed that there may have been interorganizational meetings before implementation, but little negotiation. MYWO appears to accept the terms of foreign donors, almost unequivocally, especially when they concern finances. Pathfinder put it this way, "MYWO's MCH/FP program must follow the tighter targets of USAID and submit to the restrictions on the funds . . . MYWO and Pathfinder are trapped by USAID." Pathfinder added that MYWO had more flexibility with private funds. There did not appear to be much flexibility with Coca-Cola's private funds, however.

Donors did not willingly describe the partnerships between MYWO and foreign donors and the Kenyan government. Most refused to answer the question. Of the donors who answered, all thought that there was a hierarchy between *partners* in development. The institution at the top of the hierarchy varied depending on who answered the question. For example, Pathfinder identified USAID as being at the top of the hierarchy; Marttaliitto identified MYWO as being at the top, MYWO placing itself there after the KMYWO 1989 elections as a way to demonstrate its arrogance and flex its muscles to its partners. This, Marttaliitto said, caused the breakdown of MYWO's interorganizational relations. One Peace Corps volunteer identified the Kenyan government as being at the top of the hierarchy because of its sovereignty as a state, in spite of organizational interdependencies.

One reason why donors may not have wanted to answer the question on ranking of partners within partnerships is that they may not have wanted to acknowledge the "power" they had, through the financial and other assistance they provided, relative to the government of Kenya, particularly at a time when foreign donors and the Kenyan government were butting heads over this very issue. Between 19 August 1991 and 16 April 1992, the *Daily Nation* newspaper headlines read, "Condition for aid spelt out"; "UK to link aid to human rights"; "Aid: I will not take this abuse, says Moi"; "Now Denmark halts rural project fund"; "Less aid for dictators: US sets out its conditions"; "Stop blackmailing Africa, says Salim"; "Donors: change in 6 months or else . . . "; "Conditions for German aid to developing world"; "Donor funds freeze begins to hurt Kenya—Kenya begins to feel the pinch"; "US won't aid Kenya until after elections."

Another reason why donors may not have responded to the question is that this issue of partnerships raises questions about the definition of "partner." What conditions must one meet to be a partner in a relationship? Must a partner in a relationship be "equal"? Can a "partner" ever truly be "equal"? Is it possible for one to be a "lesser" or "greater" partner? KAF stated that in its development relationship with MYWO "there was equal partnership in theory, but it was hard to live as such." KAF added, "Initially, MYWO was first among partners."

MYWO entered many partnerships as a Kenyan NGO, yet there were questions as to whether MYWO was truly an NGO. It called itself an NGO, but it was chameleon in character. It claimed an NGO status and autonomy in its development activities, to attract donors, yet between 1987 and 1991 it was officially affiliated to KANU, the ruling party, at which time it officially changed its name to KANU MYWO (KMYWO) and submitted to KANU's rules. In interviews, many donors pointed out this anomaly.

All of the foreign donors who agreed to discuss MYWO said that MYWO was not an autonomous body, not an NGO. They classified MYWO as part and parcel of KANU, both during its affiliation and after its disaffiliation with KANU. KAF was most open in its criticism of MYWO with regard to this matter. For instance, KAF was very critical of MYWO's affiliation to KANU and made charges that MYWO was KANU's cheapest hunt, calling the affiliation "dirty politics" and an attempt to control and take away the power of the women.

The majority of the donors, including KAF, said that the affiliation did not affect their decision to continue to fund and implement projects and programs with MYWO. It appeared that being able to utilize the national network of MYWO for the implementation of development projects/programs (i.e. humanitarian, environmental, diplomatic, political) was more important than MYWO's dubious status as an NGO. Marttaliitto was the exception. It indicated that it could not continue to fund MYWO when it was affiliated to KANU because of its home government's regulations. The Finnish government would not support an organization that was not a bona fide NGO. The Ford Foundation made the same argument in an unofficial interview in 1989.

Only two of the foreign donors—GTZ and KAF—indicated that they felt that their programs/projects with MYWO were successful, yet they did not attribute this success to MYWO's participation in the partnership. On the contrary, they saw MYWO national headquarters as causing hindrances for successes in development endeavors. They felt that key organizational problems of inefficiency, poor organizational performance, leadership incompetence, lack of administrative and management capabilities, lack of organizational transparency and MYWO's affiliation to and lack of autonomy from KANU were barriers to project/program successes. These problems instead led to program and project failures.

A critical issue to be considered here is how success and failure were defined and measured by the foreign donors. All of the foreign donors involved with the implementation of family planning had very vague, amorphous definitions of "success." For instance, USAID did not use the word "success," but did say that the MCH/FP project "has done well." The other donors had more obvious, tangible definitions which matched with the objectives of their projects. Some donors had more simplistic definitions of success. For instance, KAF counted MYWO's turning in financial reports on time as an indicator of success. Individual Peace Corps volunteers seemed to have the lowest thresholds for project success. Their indicators included: 1) groups setting goals, and 2) groups and individuals learning basic skills of record-keeping and marketing. It seemed that the larger the foreign donor and the more assistance it provided to MYWO, the less clear its definition of success.

With regard to measuring the success of projects/programs, MYWO as an organization had not conducted regular and consistent evaluations, if any at all. Most of the donors had not conducted evaluations either, as some did not think that evaluations were important or would have made a difference in the overall operation of the project/program. It was my impression from the foreign donors that, in the field, evaluations are not as important as those of us in the academy think they are. It was further my impression that evaluations are not considered a part of the overall development policy process. Evaluation was seen as something separate and apart from the development process. The family planning donors especially did not seem to prioritize evaluations of their projects/programs with MYWO very highly. Perhaps they suspected that evaluations would have been poor.

For the creation of more effective partnerships, disaffiliation of MYWO from KANU did not appear to most donors to be a necessary or sufficient condition. The proposed NGO Coordination Act was not seen as a sufficient condition for more effective partnerships to the three foreign donors asked. What was necessary for more effective development partnerships instead, they indicated, was increased institutional competence and accountability on MYWO's part, and greater complementarity between the partners in the interorganizational development relationships.

In sum, MYWO as the focal organization in this development cooperative partnership lacked the financial, technical and material resources to implement its programs and projects on its own, at the level that the donors provided. The foreign donors in MYWO's relational environment lacked organizational access in Kenya and clientele to serve on their own, hence the creation of the partnership with MYWO. MYWO and its foreign donors in its relational environment engaged in exchanges each to meet its own goals—MYWO to give the appearance of providing services to women and children in Kenya, and foreign donors to implement their humanitarian, environmental, diplomatic and political programs/projects.

It should be noted here that the donors who were most willing to participate in the interviews were the ones with little or no active involvement with MYWO. It appeared that the greater the extent of the foreign donor's involvement with MYWO, the less willing the donor was to participate in the interview. Often, when they did participate in the interviews, many questions regarding their feelings about MYWO or their relationship with MYWO were left unanswered. For example, GTZ flatly refused to talk about MYWO.

The reasons for the overall negative foreign donors' reactions to MYWO seemed to be the following:

1 MYWO's overtly politicized nature since its affiliation to KANU in 1987;
2 Mrs Onsando's bogus election to the chairship of MYWO in 1989;
3 The prevalent culture of fear and silence, 1991–2;
4 Mutual suspicions and hostility among interorganizational partners as well as high-level government repression; and,
5 The evolution of MYWO into a despised institution because of its ties to KANU and state repressive politics.

Kenyan government ministry bureaucrats/personnel

The following is a summary of the interviews with 19 bureaucrats of Kenyan government ministries who worked with development projects. The ministries they represent are: Agriculture, Livestock Development, Culture and Social Services (MCSS) and Home Affairs and Heritage (MHAH), which are described in Appendix D. Most of these interviewees who worked at the sub-national level would not acknowledge the presence of MYWO members within the groups with which they worked, for they insisted that they worked with the groups, *in spite* of some of the women's connections to MYWO. Two of the interviewees in this group were government special appointments. Although all of these ministries are obviously organizational extensions of the Kenyan government, many bureaucrats did not feel that they were part of the government, and made very clear the distinction they drew between the ministries' work and the government, seeing the latter as particularly Moi and the KANU party. They saw themselves as separate and apart from government. According to the interviews, their work as civil servants was for the development of Kenya, and not necessarily complementary to any work the government might be doing. Moreover, their commitment to their work was certainly not because of allegiance to Moi and KANU. There seemed to be a strong desire from many of the interviewees to disassociate from the government and the party. Many of the ministries' bureaucrats indicated that they were university graduates who were placed, upon graduation, in the Civil Service in their respective workplaces, usually away from their home areas. Two interviewees acknowledged openly, unlike most, that they were government employees.

The ministries had diverse types of relationships with their partners in development—foreign donors and MYWO. For example, NCPD had a formally established relationship with foreign donors USAID, the World Bank and CEDPA, and with MYWO national office. Together, these partners collaborated on a national MCH/FP program which was implemented in seven project areas throughout Kenya. Although NCPD was part of the MHAH, it appeared more as part of the Ministry of Health, because it worked with family planning and health projects.

The Ministry of Agriculture at the national level has ties to the foreign donor GTZ, with which it participates in an inter-ministerial approach to implement jiko-related activities. GTZ refers to the ministry not as a partner but as an active participant in the project. The interviewees from the Ministry of Agriculture, both from the national and the local levels, felt that they had unofficial ties to MYWO. Many further stated that they had "undesired" ties to MYWO. Generally, it seemed that the ministries had no official communication with MYWO national headquarters with regard to the SEP–Jiko. One interviewee said that she felt that "MYWO was nothing." Others stated clearly: "MYWO had little to do with the MYWO–Jiko." All of their cooperation—ministry with MYWO—was indirect, through GTZ. However, the ministry bureaucrats often worked with women who were individual members of MYWO, when they were in other women's groups. The relational environment within which the ministry, foreign donor and MYWO operated in this instance was very tense. It appeared that MYWO was overlooked, ignored and dismissed as the focal organization in this particular interorganizational relationship.

The Ministry of Agriculture seemed more closely associated with the MYWO–Jiko project as the interviews suggested, but the SEP–Jiko project was officially a project of the Ministry of Energy.

One interviewee officially cooperated with the foreign donor DANIDA (which had no direct ties to MYWO) and a local women's group. Some MYWO members were members of this women's group and were in the MYWO's local leadership as well. This cooperative project was not coordinated through MYWO national or local office. All MYWO connections were coincidental.

This same interviewee saw MYWO somewhat differently from the other interviewees. She was a local MYWO leader, in her capacity as a MYWO chair and simultaneously as the chair of the District Development Committee (DDC). She felt that her overlapping roles were positive because, in them concurrently, she was able to politically engineer benefits for the women's group in which she was a member. She felt that her various power bases had been critical for the group's success. Though she talked of MYWO within the locality in which she worked, she did not address MYWO at the national level.

MCSS had no formal projects with MYWO. In some areas, there were high levels of interaction between MCSS and MYWO, and in some areas there were not. In some areas, there was assistance from foreign donors—locally based assistance, not secured through MYWO national headquarters. In a greater number of areas, however, there was not assistance from foreign donors. Much of the cooperation was with local ministries, and much of the assistance was locally based. Foreign donors who funded rural groups seemed to depend on the reputation of the local women's groups in determining whether or not they would assist them.

Another interviewee worked with MYWO at the national level. From this vantage point, she stated that part of MYWO's problems, and a reason for MYWO not having much success as an organization, is that MYWO has no national centralized planning. Another reason she cited is that it was impossible to distinguish between MYWO's and MCSS's groups, making isolation and evaluation of MYWO projects near impossible. Further complicating matters, she stated that there were major divisions in MYWO at the national level between the program managers and the elected officials. On one hand, the national program managers were professionals who were concerned that programs were implemented, and they worked toward that end. On the other hand, the elected officers were inexperienced in program implementation, and were more concerned about politics than about implementation of programs for development. From her observations and cursory evaluations, she stated that MYWO did not (and perhaps could not) have an impact at the village and/or grassroots level.

MYWO and its national elected officials were despised by most of the ministries' interviewees. Questions about MYWO nearly ruined many of my opportunities to conduct interviews. Many interviewees simply did not want to talk about MYWO. MYWO's unpopularity with the ministries' interviewees was resounding. From the interviews, there seemed to be two main reasons for this: 1) many people felt that MYWO did not do much development work, and that they merely got the credit for the work others—including the ministries—had done; and 2) many felt that since affiliation with KANU, MYWO had become "big-headed." For example, one of the interviewees described MYWO's behavior as equivalent to "haughty," for with their "partners" it was their stance that "If you don't do this [follow MYWO's orders], you'll know that we are KANU." The consensus from the interviewees was the same as that of the partners in development: the national elected officials of MYWO who represented MYWO did not have a cooperative spirit. It was also expressed by some that MYWO's destiny was inextricably tied to KANU. They did not want to be associated with this. Many saw KANU as a sinking ship.

It was obvious from the interviewees' responses to questions that they did not consider MYWO an autonomous or indigenous organization. One interviewee was most expressive about this. She stated, "MYWO does not have an indigenous agenda, and they [MYWO leaders] have little resistance to donors' agendas. When the women were left alone, they were indigenous. They have lost their direction with the affiliation to KANU."

Some of the other interviewees further stated that the ministries' relationship with MYWO became confused when MYWO was affiliated to KANU, and when the 1989 elections were held. Although local MYWO representatives were elected, they had no network or effective avenue to communicate with MYWO at the national level. As one interviewee stated, "Coordination is there at the local level, but not at the national level. Local level MYWO needs more contact with national level MYWO. The problem comes from the national level." Clearly, this caused a bottleneck, for there could be no reliable back and forth communication between local and national leaders, hence local MYWO leaders could not be accountable to the local women. This caused resentment from the local women toward local MYWO leaders.

Overall and somewhat paradoxically, from the responses of these interviewees it seems that project coordination worked better when MYWO at the national headquarters was *not* involved. Stated differently, project coordination seemed to work better when local ministry personnel had access to local donors, and when they did not have to go through the MYWO national office. To the few interviewees who were willing to discuss MYWO's disaffiliation from KANU, disaffiliation seemed to offer little increased hope that MYWO at the national headquarters would not continue to stand in the way of interorganizational cooperation. Disaffiliation was seen as a symbolic act for foreign donors' continued investment in the development of Kenya. One interviewee explained it this way, to paraphrase: disaffiliation was action on which MYWO had to follow through to appease foreign donors and stay in their good graces in order to continue to get funding. MYWO will not effectively disaffiliate from KANU, however, since its key elected officials and appointees, including the chair and CEO, are KANU members themselves. If the women are not active members of KANU, they are the wives of KANU members.

Almost half of the interviewees stated that they did not regard MYWO as a critical actor in development partnerships. One interviewee said that MYWO *should* be a critical actor, because it *should* represent all women and all women's interests in Kenya. However, NCPD stated that MYWO is a critical partner in the MCH/FP program, where an unstable hierarchy existed because MYWO and foreign donors each waged threats to get the program implemented according to "its agenda"—MYWO pushing its KANU weight around and foreign donors threatening to take back their money. Another interviewee viewed this differently. She identified the government of Kenya and its National Development Plan at the top of the hierarchy. Another interviewee at MYWO offered important insight in this matter. She stated that MYWO was important for development partnerships, because it could be the implementing arm of a partnership. In terms of program and project formulation, however, she admitted that MYWO had no power. She stipulated that it is the government and the foreign donors—in shifting positions—who have power in policy/program and project formulation. The impact that this has on the hierarchical ordering of a partnership, she argued, is that there is not "one" consistent partner at the apex all of the time. Instead, there is a shifting leadership position for the top of the hierarchy between government and foreign donors. Who is at the apex depends on what is at issue—policy compliance or resource necessity. In all negotiations, however, she stated, MYWO is consistently in the weakest position. She used the incident between NORAD, Wamwere and the government of Kenya to illustrate this point. She further demonstrated the strength of foreign donors and the weakness of MYWO by stating that an approximate breakdown of resource contributions reveals foreign donor contributions to MYWO to be 70 percent of its overall operating resources (9 percent short of MYWO's staff person's figures), while KANU contributions are 20 percent, and MYWO's 10 percent. Because of this breakdown, she stated that donors tend to force what they want on NGOs (in which she included MYWO). Foreign donors are able to do this because they have the access to Kenyan political elites that Kenyans themselves do not have.

In terms of how this resource contribution affects MYWO agenda changes, the interviewee stated that, generally, it seemed that when program and project changes were proposed by a partner in the development partnerships, meetings were held to determine if and what type of changes should be made. She recounted that when the proposed changes were about resources, the foreign donor usually prevailed; when the proposed changes were about policy, the Kenyan government usually prevailed. Basically, foreign donors were unbending about resource issues, and the Kenyan government was unbending about policy issues.

Most of the interviewees felt that a significant number of their projects were successes. The bureaucrats from the Ministries of Agriculture and NCPD were more definitive in their answers since they had objectives, goals and numerical targets which were tangible and could be measured.

MCSS's responses regarding goals were different. They spoke of broader, more fluid goals, such as assisting groups with problems, listening to their problems, assisting them in finding solutions and looking for markets for their goods. MCSS seemed less focused in their objectives than the other interviewees. A reason for this might be that the bureaucrats in the MCSS are not trained in technical fields. That is, the personnel in the MCSS are social workers by training, who may possess empathy, altruism and social vision, but unlike the personnel from the Ministry of Agriculture Home Economics Extension, they had not learned technical skills which they could transfer to women's groups. MCSS seemed to serve more of a motivational and support purpose for the groups. The MCSS interviewees, when compared to other ministry interviewees, seemed more likely to link their definition of project success to whether or not a project or group had been able to secure a foreign donor.

One interviewee provided the most critical view of successes and failures of MYWO programs. Overall, she felt that MYWO programs had not achieved a lot of success. The most successful (which was about 60 percent successful in her estimation) was the SEP–Jiko program because of its education, outreach and acceptance among women. The LD program was not very practical, she believed, and therefore not very successful. The MCH/FP was also not very successful, in that it served a few women at very high costs and was limited to certain women in certain ethnic groups and areas. Moreover, she pointed out that MYWO did not own its own clinics and its budgets were very prohibitive because of interorganizational stipulations. The Nutrition program had ended when she began her appointment with MYWO, so her knowledge of this program was very limited. The Coca-Cola program had not yet begun its implementation.

The majority of the interviewees agreed that it was the cooperation, albeit limited, between partners—foreign donors and ministries primarily—that created the development successes. MYWO was not identified by any of the interviewees, except NCPD, as a reason for program success. NCPD did qualify its position, however, by stating that its cooperation with MYWO was difficult at best. Other institutions the interviewees credited for development successes are: other women's groups, local chiefs and some individual women leaders, especially Jane Kiano, Joan Mjomba and Wangari Maathai. Many of the interviewees felt that MYWO was actually a hindrance to success.

MYWO's relationship with the government and foreign donors was seen as having no significant positive impact on gender equality. For the most part, gender relations were regarded as staying quite the same.

The interviewees made many suggestions for more effective approaches to development and development partnerships:

1 NCPD said that it (NCPD) should be strengthened as an institution and that its ministry workers should be paid more as an incentive to do their jobs. Moreover, it said that contraceptives should be sold as a means to generate income. About MYWO, one interviewee only commented, "I'm not so interested [in MYWO] any more since it is KANU. It's bad considering how people feel about KANU these days."
2 The Ministry of Agriculture interviewees suggested the following: (a) More needs assessments should be conducted; (b) No impositions of projects from outside should be allowed; (c) No technical experts from abroad should be brought in when Kenya already has the expertise; (d) Foreign donors should make the provision of money their priority, and do

not need to be in Kenya to supervise; and (e) Groups should communicate more with ministries;

3 MCSS had fewer and less specific recommendations: (a) More coordination between partners is necessary; (b) Donors must understand the unique cultural background of each area; (c) Assistance from foreign donors should be given directly to women's groups; (d) Men should not sit on the executive of any women's groups; and (e) Men should allow women to work without fear.

4 The Ministry of Livestock Development indicated that women's groups should solicit their assistance at the beginning of their projects, not when it is too late to save the livestock and salvage the project;

5 A final interviewee recommended that: (a) MYWO should engage in more central planning to create development plans which are continuous; (b) MYWO should correct its leadership problems by having new and honest elections; (c) MYWO should identify and use local resources, as well as woo Kenyan businesses and professional elites; and (d) the Kenyan government should leave women alone to find their direction again and reindigenize.

Other relevant interviewees

The following is a summary of interviews and discussions I had with 21 knowledgeable persons regarding the politics of development cooperation. These interviewees are an unintended set of relevant persons who were not initially part of my intended interview groups. Due to unanticipated circumstances in the field they became a very important resource and data base. A general description of these 21 persons was provided in Chapter 4.

A very different view of MYWO was presented by this set of interviews as compared to the other interviews, and a more consistently critical perspective of MYWO's "partners" in development was presented as well. Many of the interviews described MYWO as a middle-class women's organization that never really caught on at the grassroots, even after the organization disaffiliated from the colonial state and was Africanized in 1961. MYWO's evolution out of the colonial period, having been started by the wives of colonial officers, seems to have branded the organization *kikoloni* (colonial), and this has created a reputation for MYWO which would come back to haunt the organization only 25 years after Kenya's independence. That is, MYWO had again become an official vessel and an agent of the state in 1987, affiliated not with the British colonial state but to the Kenyan *independent* state.

With regard to the affiliation of MYWO to KANU, one of the interviewees recounted the following: since the 1970s, there had been a trend to destroy organizations in Kenya, especially those that were land-buying, and especially women's organizations that functioned as interest groups. This trend, she said, was reflective of an overall government attempt to disintegrate civil society. It seemed that when the government could not control an entity it set out to destroy it, especially if that entity was forceful enough to influence the politics of the country. Thus in 1976, when women at the grassroots showed national and international development potential, the government created the Women's Bureau as a component of the Ministry of Culture and Social Services, to control women's activities and undermine their political potential.

Just over one decade later, in 1987, the interviewee continued, MYWO refused to submit to coordination with the Women's Bureau under Jane Kiano, and was affiliated to KANU. Two years after that, in 1989, MYWO was named, under Moi's mandate, the functionary for channeling financial and other assistance to all women's groups and women's projects in Kenya. Hence, many foreign donors duly regarded MYWO as the central mechanism for distributing financial assistance to women in Kenya, though many of them began to question MYWO's legitimacy in this capacity. Another interviewee said that KANU had affiliated MYWO not only because MYWO represented a vast number of votes and political support, but also because MYWO was considered one of the easier organizations to co-opt and control. That is, KANU considered women to be weaker than men, and thus easier to control. The interviewee compared MYWO to the Central Organization of Trade Unions (COTU) which is predominately male and which KANU had tried, but failed, to affiliate.

Many of the interviewees felt that KANU affiliated MYWO purely for political reasons. KANU wanted to control MYWO's activities and development agenda, as well as covet the money MYWO was able to secure for development purposes. One interviewee described MYWO as merely a tea club, existing solely for the exploitation of KANU men.

Foreign donors were also seen by many of the interviewees as *not* merely providing financial and other assistance to MYWO with no strings attached. They were considered a devious lot. Most of the interviewees were of the opinion that MYWO's agenda is that of its Northern donors, since MYWO accepts most of its finances from the North. The interviewees felt that MYWO really had no room to negotiate with its foreign donors, as they argued that it is the donors and not MYWO keeping the organization in operation financially. One of the interviewees said that it was not the fact that MYWO had foreign donors that was the problem. The problem was that MYWO did not have a diversity of donors, each providing a small amount of funding. Instead MYWO had only a few donors who provided the bulk of its finances. The interviewee who took this view (herself a director of a successful NGO in partnership with donors) argued that an organization utilizing outside funding must secure a diverse donor base to maintain its agenda and its autonomy. She said that many Kenyan NGOs can be autonomous, though many are not. The key to maintaining autonomy, she said, is having a diversity of funding sources and a clear, defined agenda *before* the money is accepted. She further added that an NGO must know what it is doing, and it must also know what the donor is doing. The interviewee stressed that money should not be taken just because it is offered. One must be willing and able to say no, when the terms are not acceptable.

It was expressed by most of the interviewees that foreign donors use their money to bring clout and influence into Kenya. Moreover, they said that foreign donors were not interested in MYWO's or other Kenyan organizations' agendas because they "want to do what they want to do," and Kenya puts up no effort to negotiate or resist. One interviewee, for example, stated that "Kenya's National Development Plan merely mimics the World Bank." It was also added by many interviewees that Kenya's exploitation by foreign donors was obvious through such requests by the IMF and World Bank to cut the Kenyan Civil Service by 25 percent and teachers by 50 percent, which amounts to millions of people. Otherwise, Kenya would lose its external assistance, as reported in the 31 January 1994 issue of the *Daily Nation*.

Many interviewees were very critical of the in-country expatriate staffs of the foreign donors. They questioned their commitment to Kenya's development. They claimed that many of the staff knew little about Kenya and Kenyans, and that many were not genuinely interested in learning. Instead, they claimed, the in-country expatriate staffs were more interested in "big salaries, taking safaris to national parks and mingling with other expatriates." They further commented on seeing the same faces around town forever, implying that there is a small network of expatriates who make up an "elite international donor community" getting more than they give by working in development in Africa. Many interviewees believed that most expatriates did not take development seriously; those who did saw development as their "job." One interviewee explained that, in order to be effective, "One must live development, not just do it as a job."

Several of the interviewees felt that, as Kenyans, their hands were tied with regard to development in their country. They felt that they could do nothing to influence foreign donor agendas and, by extension, Kenya's development agenda. They claimed that foreign donors had more say in Kenya than they (Kenyan citizens) did, and that foreign donors had access to networks of political elites that the average Kenyan citizen did not have. They alleged that foreign donors were aware of this and utilized it to push their "development" agendas, whether Kenyans agreed with them or not. This, they referred to as "diplomatic arrogance." These interviewees had little hope that their political representatives or Kenyan government ministries would change the situation. One interviewee stated it in this way: "Politicians do not understand the process of development. They do not understand the necessity of needs assessments and studies and they often stand in the way of development. They want to use [development] money for huge structures and vehicles."

Many interviewees also alleged that politicians use development money to line their pockets, and that the "ministries were lax and that of all the ministries, MCSS is the weakest. There is no discipline." The interviewees did not make a distinction between the KANU government and ministries. Overall, these interviewees felt that there was too much of a focus on money and not enough of a focus on development policy negotiations, technical training and credit schemes versus aid.

MYWO was not seen as an entity to enact change either, as it is part of the state, bent on maintaining the status quo. Instead, it was seen, under the leadership of Mrs Onsando, as directing much of its attention and money to the Kisii area, the area from where Mrs Onsando originates. Thus, in addition to charges by interviewees of being middle-class political pimps, MYWO elected officials were also charged with being "tribalistic."

These are some of the reasons why some foreign donors refused to discuss MYWO at length. The Japanese International Cooperation Agency (JICA), who conducted surveys of the needs of women in Kenya in early 1992, decided not to support MYWO in any of its current programs or begin a new program in conjunction with them, because of MYWO's organizational affiliation with KANU and because of the foci of MYWO's programs. DANIDA refused to discuss MYWO as well, merely saying MYWO had been "bulldozed" by KANU and no one was interested in them any longer. To put an end to my questions and probing about MYWO, the DANIDA interviewee very curtly said, "We are talking about MYWO, not about politics." As an employee of DANIDA she was more critical of DANIDA, using rhetoric about women but not really having a Women in Development (WID) component, as well as not really being sensitized and committed to women in general.

MYWO was clearly more popular outside Kenya than it was inside the country. Many interviewees clearly stated outright that it is the women of Kenya who keep the economy going, but many of these women, it seems, operate outside of MYWO, even though MYWO (mis)represents itself as the umbrella organization of all women's groups in Kenya. Its reputation is very different from reality.

Many women in their twenties and thirties, especially those who live in Nairobi, are not interested in joining MYWO at all. Many feel that MYWO is not a viable organization for women any more, because of its corrupt nature, affiliation to KANU and colonial legacy. Instead these younger women are drawn to more professional organizations, local organizations in their area and, increasingly, private entrepreneurship. For example, one woman interviewee was a member of five women's groups in her area, in which all groups used the merry-go-round concept. In a merry-go-round group, each group member contributes a set amount of money weekly and one by one each member receives the lump sum contribution at the group's weekly meeting. If there is a group of ten members who each contribute Ksh 20 each week for ten weeks, each member will receive Ksh 200 at her turn in the merry-go-round. In addition, professional groups have increasing numbers of younger women members. For instance, the Kenya Professional and Business Women's Club (KPBWC), under the leadership of Beth Mugo, has been gaining tremendous momentum as Mugo encourages women to "go it alone" in business endeavors. There are also many younger women not interested in women's organizations at all; instead they are seriously undertaking individual business enterprises.

With regard to the overall inefficiency of MYWO in turning in reports and records to foreign donors, accurately and on time, one of the younger women interviewed said, "Not to worry, the inefficiency of MYWO is intentional. If the records are not done properly, they must be redone. That means that the person responsible can keep [her] job and pay longer, because there is still work to do."

In sum, the interviewees seemed fed up with Moi, KANU, and MYWO. Though many of them wanted change and pushed for change, they were not very enthusiastic that change would come with multiparty politics and impending elections. Just as MYWO's disaffiliation from KANU was seen as symbolic and bogus, so were the impending presidential and parliamentary elections. The interviewees seemed lost as to what to do about corruption and repression. It seemed to be a situation in which one would either accept it and hope to personally benefit from it, or despair. Some interviewees in the NGO sector, however, had some recourse. They had organized an NGO Task Force for the Eastern and Southern Africa regions to address the growing frustration of finding themselves, as they argued, "responding increasingly to donor's priorities rather than to those of the communities they serve." One way that they were addressing this was to challenge the NGO Coordination Act of 1990, which was created to register and license all NGOs—international, national and local—operating in Kenya.

The following chapter will discuss the findings presented in this chapter as they relate to the politics of development cooperation. It will also present an alternative model for guiding inquiries into development interorganizational relationships involving NGOs, women, foreign donors and African governments.

6

THE WEB OF DECEIT: GRASSROOTS DEVELOPMENT CAUGHT?

The previous chapter demonstrated a barrage of contradictions about MYWO and its relationships. Some of these contradictions exist because there is confusion about what MYWO is actually doing in terms of organizational and interorganizational activities. Some of these contradictions reflect the dual character of MYWO—the development organization, and the political KANU appendage. Some of the contradictions seem intentional from the MYWO leadership, attempting to cause confusion and inertia among women.

This chapter will discuss the significance of the research findings documented in the last chapter and present some conclusions about the politics of development cooperation. More specifically, Part I of this chapter will discuss the main research findings and the contributions they make to the various bodies of literature discussed in Chapter 2. It will also tease out the implications of the research findings for development cooperation relative to organizational integrity, authenticity, will, autonomy and dependence. I will also raise questions. Part II of this chapter will consider the prospects for cross-national development cooperation studies and make recommendations for development implementation in the field and for further research. It will also present some concluding remarks.

PART I: DISCUSSION OF RESEARCH FINDINGS

Of the main research findings, this research demonstrated that MYWO was, in the recent past, recognized as a development organization, more so outside Kenya, perhaps in academic circles and by some foreign donors, than it was noted as such in Kenya by Kenyans and by organizations in its relational environment. This research determined that MYWO has a falsified reputation and that it consciously misrepresents its activities and linkages to grassroots women in order to secure foreign donors. MYWO elected officials attach the organization to successful individuals, such as Mrs Mjomba and Mrs Kiano, and businesses like the family pottery business discussed in Chapter 5, to steal their reputation for MYWO. MYWO—national headquarters—also competes with other women's groups such as NCWK and KANU Women's League for preference with KANU and foreign donors. KANU purposely affiliated MYWO and made it an extension of the ruling party so as to force MYWO to compete with KANU Women's League, Nairobi. KANU projected that the party would ultimately benefit through using a "divide and conquer" strategy against the women. It did.

Women's groups

There was a definite disjuncture between being a member of a grassroots women's group and being a member of MYWO. That is, grassroots women's membership in groups seemed to be based primarily on promoting women's togetherness, and MYWO membership seemed to be grounded more in certain individual women's expectations of gaining political and economic power by exploiting women's mobilization and solidarity. These individual women are MYWO national leaders who attempted to become rich and/or gain political recognition by endearing themselves to KANU. They did not genuinely promote or fundamentally believe in women's solidarity. They are anti-feminist.[1]

KANU realized these women's intentions, promoted their beliefs in individualism and used them. One obvious way in which it did this was by declaring MYWO the clearinghouse for all women's activities and finances in 1989 (although this did not take root). In the clearinghouse role, MYWO national leaders would continue to use the "women in development" propaganda to secure foreign financial assistance for MYWO on behalf of the women's groups. In actuality, however, the donor assistance would be channeled through MYWO, guarded by elected officials and placed at KANU's disposal. Likewise, with the new arrangements governing interorganizational relations between KANU and MYWO, after affiliation and the 1989 elections, KANU would assume authority over MYWO in determining how this assistance would be spent.

That KANU would capture MYWO's organizational finances and that MYWO national elected officials would front as agents for KANU to secure foreign donations is not surprising, since MYWO elected leaders were politicians, desperately aspiring to be recognized by KANU men. As politicians, these women undermined development endeavors. They were not development professionals; and they were not committed to grassroots development causes. Instead, MYWO national elected officials were more identified with the male politicians, whose political maneuverings and trickery had amassed fortunes for many of them. As Fatton demonstrated, women's access to power and resources in Africa is directly dependent on their linkages to men.[2] This was the case in Kenya, to the extent that if gaining power meant betraying women's causes, many women did just that. As Bujra has argued, and as this case study has demonstrated, women's organizations do not necessarily mean that its female leaders and memberships espouse or hold true to a feminist ideology.[3]

In effect, MYWO national leaders were not responsive and accountable to the women's groups that they claimed to represent. Network linkages between elected officials and women's groups were usually mythical. Hence, women—groups and individuals—were disenchanted with MYWO not only because of its political nature, but also because of its failure to link with them. Although networks which interlinked national MYWO with provinces, districts, divisions, locations, sublocations and villages had been created, particularly since the MYWO decentralized elections of 1989, interlevel communication remained elusive.

Women's groups at the grassroots had more faith in and relatively stronger (though still tenuous) linkages with the foreign donors in their home areas and with local government ministries who assisted them in their development endeavors than they did with MYWO. Women's groups' expectations of foreign donors seemed to be greater than their expectations of the Kenyan government, since many women were both knowledgeable of the limitations of the government ministries and disenchanted with KANU—party and government. Most saw MYWO not as an indigenous women's organization, but as part and parcel of KANU.

MYWO elected officials and staff

Elected officials

Among MYWO's elected officials, there was a diversity and inconsistency of organizational purposes and goals. The women in MYWO's elected leadership were not unified in their goals for the organization. Onsando was trying to change the organization into an income-generating and profit-making business, although she continued to claim that MYWO was an NGO. Onsando's attempts to turn MYWO into a business were clear, but she did not have organizational backing, nor did she have the mobilization skills to garner support for her plan. Even so, Onsando did not care if she had the backing of MYWO, for she looked to KANU (her patron) rather than to MYWO for support, direction, approval and protection.

One of the critical issues of the Onsando administration which contributed to its problems is that MYWO had no national planning. That is, the elected leaders did not set specific goals and objectives, or formulate policies or plans for program and project implementation for the organization. When the development professionals within MYWO, (e.g. the professional staff, specifically program managers) attempted to lay a foundation for continuity and perpetuity of programs, the elected officials undermined their efforts. It seemed as though the elected officials intentionally blocked program implementation and, in effect, sabotaged potential development program successes. As a result of MYWO elected officials and professional staff discontinuity, MYWO members—national staff and local women—were frustrated because they had no idea of the organization's day-to-day agenda, much less its long-range goals and direction. There was virtually no intraorganizational communication, laterally or hierarchically. Veteran MYWO members were disturbed with what they perceived to be a shift of the organization's focus from Kiano's time of "social welfare and development" to "politics and individual financial aggrandizement" under Onsando.

During the Onsando administration, MYWO's interorganizational relations, specifically regarding financial matters, were kept very secret, even from the national assistant treasurer, Mrs Mjomba.[4] Mjomba was used by MYWO for her reputation and connections, and was ignored after her usefulness had ceased to serve the other elected officials. Mjomba, as a result, tried to use resources at her disposal to get back at them.

Beyond some program managers, no one really seriously asked about the finances of the organization. This research demonstrated that women sometimes wondered about the monies; but the consensus was they were going to be lied to by MYWO leaders (and KANU) anyway, since, they argued, deceit and corruption were the norm. The women were also fearful to ask—there might be repercussions. Basically, MYWO was not accountable to women's groups and MYWO members, and KANU was not accountable to MYWO. MYWO was far from transparent and, overall, MYWO's interorganizational relations lacked accountability and transparency. In addition, MYWO was a closed organization, very vulnerable to graft.

In MYWO's intraorganizational relations, MYWO elected officials were disconnected from grassroots women; and, they were fierce opponents of MYWO's professional staff. In MYWO's interorganizational relations, MYWO elected officials were controlled and manipulated by KANU; and MYWO was schizophrenic towards foreign donors.

In interorganizational relations with foreign donors, MYWO elected officials spent most of their energy either courting them for aid or criticizing their restrictions on aid, often simultaneously. MYWO likened their restrictions on aid to the imposition of foreign ideals—which in 1991–2 were liberalization of the Kenyan economy and the introduction of multiparty politics. After foreign donors publicly tied economic liberalization and political reform to aid in November 1991, they were constantly at the center of government and MYWO criticism—as MYWO was the KANU government's mouthpiece. Foreign donors were blamed for everything from failed projects to MYWO's and KANU's poor performance reputations, to donor withdrawals, to inciting local people towards popular unrest and demands for multipartyism. Ironically, MYWO engaged in no self-criticism or self-evaluation, nor did it take a critical look at the KANU party and government with regard to the issues aforementioned.

One of the factors which led to the schizophrenic posture that MYWO elected officials took vis-à-vis foreign donors is that instead of focusing on program development and sustainability for the organization, MYWO elected officials focused more on "getting money" from donors. As a result they did not lay a solid programmatic foundation for the organization. Instead, they uncritically embraced donors' programs and projects as long as donors "brought money." This research found that MYWO had no current grassroots projects. MYWO projects were basically extensions of government projects, or foreign donors' projects, or residuals of Kiano's projects, or a combination thereof. The motto of the elected officials in most instances was "get the money and then worry about the project." As a result, donors did not have to do much "imposing" of their "foreign" development agendas in their interorganizational relations with MYWO. MYWO, and by extension KANU, showed virtually no resistance, as long as donors brought money.

MYWO's schizophrenic behavior was obvious in its shifting positions in its interorganizational network. MYWO did not operate in the static manner that Jonsson describes of a focal or linking pin organization in the "game theoretical notion of interdependence."[5] Instead MYWO was in constant flux. To potential foreign donors, MYWO misrepresented itself as an indigenous women's NGO leading development in Kenya, and thus it took a position at the top of the development hierarchy, directing its partners. For KANU's benefit, however, MYWO had to take both the lead position in interorganizational relations, to secure monies from foreign donors, and also a subservient position to KANU simultaneously, to demonstrate that it knew its "rightful place." As a women's organization, it was MYWO's responsibility to take a subservient position to KANU men, as politically, economically, socially and culturally women could *not* be equal to men. Thus in interorganizational relations, because of gender differences MYWO could *not* really be an equal partner to KANU, although it had to give the impression to foreign donors that it could.

The female private sphere and male public sphere dichotomy—previously discussed in Chapter 2 and which Scott and Nzomo each discussed in relation to women and development—argues that social, political and economic life are indeed gendered. That is, the public sphere of politics and economics remains overwhelmingly, both effectively and symbolically, the domain of men; and the private sphere of the home, domesticity and childcare remains overwhelmingly, both effectively and symbolically, the domain of women, despite the increasing participation of women in public life. This was in fact the case with MYWO and KANU. Moreover, some MYWO elected officials rationalized their subservient position, and justified KANU's generally denigrating, patronizing and misogynistic statements and actions. Despite (limited) male and female interactions in public life, men and women live separate realities.

Staff

Just as MYWO national headquarters was not interorganizationally linked to women's groups at the grassroots, MYWO program managers (i.e. MYWO professional staff) at the national level were not very closely linked to MYWO elected officials either. Whereas elected officials were more closely linked to KANU than to the staff, the staff had closer ties to the grassroots than to the elected officials and KANU. That is, the staff was only closely linked to the grassroots in those areas where their programs were operational.

The national MYWO staff seemed also to have stronger linkages to the foreign donors than it did to the Kenyan government, even though some of MYWO's programs are appendages of government programs. In spite of this, there was generally little acknowledgment of the role of Kenyan government ministries in, or their contributions to, development partnerships. MYWO staff relied much more, both financially and psychologically, on foreign donors for project support and implementation. Moreover, MYWO staff seemed to work more closely with foreign donors than with MYWO elected officials. Their psychological reliance on foreign donors seemed so great that it explained, to some extent, why there was little belief at MYWO headquarters in project sustainability if donors withdrew.[6]

Some national program managers, in their capacity as development professionals, were very frustrated with MYWO's politics in its interorganizational relations. Moreover, they were anxious about the funding for their programs—because funding uncertainty was problematic. On one hand, some of the MYWO staff took an apolitical stance, trying to keep politics, development and personal feelings separate. On the other hand, however, some of the staff were activists who linked politics, development and personal criticism. That is, they took stances against MYWO elected officials and KANU for deceiving women and undermining MYWO's development potential. They blamed KANU for placing a group of incompetent elected officials at the helm of MYWO. They argued that these women did not possess the most rudimentary of skills to run the organization. They blamed these women, and by extension KANU, for MYWO's poor performance in intra- and interorganizational relations. This ignited a great deal of friction and animosity between the two groups—elected officials and program managers, particularly at the national level. Elected officials had no proof which would exonerate them from the charges against them.

Intraorganizational relations were somewhat different at the subnational levels. Generally at these levels program staff were not very knowledgeable about MYWO national headquarters activities with its partners. Elected officials and staff worked together more harmoniously, though not always. Although the subnational levels looked to the national headquarters for direction, national headquarters could not provide any because national elected officials had no development direction or clear organizational goals. Moreover, MYWO elected officials at the headquarters did more to block networks for activities and communication than they did to facilitate them.

With regard to interorganizational partnerships, the national program managers readily acknowledged that the government was at the head of the development partnerships. They had no illusions that women led these partnerships. They acknowledged that hierarchical relations existed, instead of equality between partners. They differed, however, in that some saw MYWO relations with government as apolitical, and they justified the government's position at the head of the partnerships. They argued that the government merely served as a watchdog of Kenya's development, intervening only on policy matters to ensure the implementation of Kenya's "master plan" for development. Others did not justify the government's maneuvering towards the apex of development partnerships. They saw the government as basically conspiring to 1) control development partnerships, and 2) maintain political, economic and patriarchal power as well as foreign assistance at all costs. They argued that the government was not really concerned about grassroots development or women.

Although most of the staff readily acknowledged that MYWO was not autonomous from KANU, some program managers tried to keep up appearances that it was. They tried to rationalize MYWO's autonomy within a merger. They also did a cheerleading routine about the success of their projects. This "successfulness" and "autonomy" propaganda proved to be strictly for the benefit of the foreign donors in order to continue to get funding. Program managers and other staff were fully aware of MYWO's programs and projects limitations and the organization's relative overall lack of success and ineffectiveness, but they were not likely to reveal this if it meant donors would be lost. One interviewee in particular was most categorical in affirming that MYWO could not be autonomous from KANU and USAID.

Foreign donors

Regarding MYWO's interorganizational relationship with foreign donors, it is critical to note the diversity of the international donors involved in IOR relationships. They are a diverse lot, ranging from international governmental organizations to businesses. Despite their differences, foreign donors providing assistance to MYWO tended to be lumped into a generic category of "NGOs," not by scholars, but by persons involved in the planning and implementation of projects and programs in the development milieu in Kenya. This generic categorization of all donors as NGOs is problematic for research for several reasons: 1) it challenges determining exactly what this

NGO phenomenon really is; 2) it complicates unraveling the kinds of relationships that develop and exist within development partnerships of "NGOs" and their relational environment; and 3) it does not facilitate assessing future IOR development prospects in which "NGOs" remain critical partners in development cooperative endeavors. MYWO, to some, has been consistently regarded as an NGO. It has also changed its face to appear as an NGO when it maximized the benefits it could get from foreign donors. These benefits, however, often did not trickle down *intra* organizationally.

Ironically, none of the foreign donors that funded MYWO development programs and projects are technically non-governmental organizations (NGOs) according to the working definition of NGO provided in Chapter 2. This raises questions as to the amorphousness of this NGO phenomenon. See Table 6.1 for the (re)classification of foreign donors. Coca-Cola, for example, is a private business enterprise as discussed in Chapter 2. It is not an NGO because it operates primarily for profit. In addition, none of the other foreign donors were significantly disassociated from their home governments in terms of funding, and therefore they are not genuine NGOs either. Moreover, many of their development agendas, including monitoring, were also closely tied to their home governments, again eluding the NGO definition.

For example, USAID, the Peace Corps and the World Bank are all direct governmental organizations. That is, USAID, created in 1961 by the Foreign Assistance Act, is the primary economic assistance agency of the US government which implements broad foreign policy objectives of the US in the developing world.[7] The Peace Corps, also created in 1961 by the Peace Corps Act, is a US technical assistance organization created to foster skill transfers and cultural exchange between the US and the developing world.[8] The World Bank is a specialized agency of the UN and an international organization created at the Bretton Woods Conference in 1944 to assist in the reconstruction and development of member territories. Today, the World Bank's focus, as its rhetoric stipulates, is on providing economic assistance to

Table 6.1 Foreign donors funding MYWO programs/projects

Governmental organizations (GOs)	*International governmental organizations (IGOs)*	*Quais non-governmental organizations (QNGOs)*	*Non-governmental organizations (NGOs)*	*Business*
USAID	World Bank	Pathfinder International		Coca-Cola
Peace Corps		CEDPA GTZ KAF Marttaliitto		

developing countries.[9] The World Bank, USAID and the Peace Corps are classified as IGO, GO and GO respectively, as Table 6.1 illustrates.

Pathfinder International claims to be a private philanthropic organization—an NGO. Since the late 1980s, Pathfinder has received over 90 percent of its funding from USAID, and it also implements USAID policies.[10] Hence, Pathfinder is quasi-governmental. CEDPA has not provided information on its funding and agenda matters, yet previous studies as well as my field interviews indicate that CEDPA relies heavily on USAID funds and USAID's family planning program agenda. Its agenda reportedly changes according to US congressional modifications of financial allocations for family planning programs abroad.[11] CEDPA also is quasi-governmental.

German-based organizations, GTZ and KAF, are quasi-governmental as well. GTZ is both a business enterprise and a public corporation. It works directly with the German Federal Ministry for Economic Cooperation and Development in the scope of the German federal government's development policy and is subject to government regulation.[12] KAF receives most of its funding from the German Foreign Ministry, the Ministry of the Interior and the Ministry for International Cooperation. It receives hardly any private contributions. Although KAF claims not to be tied to the German government or the German Christian Democratic Movement of which Konrad Adenauer, the first Chancellor of the West German government, was a principal founder, its activities and fiscal operations today are monitored by the German government.[13]

Marttaliitto of Finland is also quasi-governmental. It receives 75 percent of its financing from the Finnish government and is subject to government regulations. In Finland, Marttaliitto is regarded as a state agency.[14]

All of these organizations with the exception of Coca-Cola depend substantially on their home governments for their operating funds, and four of them—USAID, the Peace Corps, the World Bank and Pathfinder—are directly beholden to their home and member governments in pursuance of their objectives. These organizations are part and parcel of state and interstate governing apparatuses in that they are relatively permanent and powerful institutions which establish and enforce policy outlines and priorities.

With regard to agenda adoption and implementation, USAID adopts and implements policies of the International Development Cooperation Agency (IDCA) whose specific charge is "policy planning, policy making, and policy coordination . . . through the effective use of US bilateral development assistance programs and US participation in multilateral development organizations."[15] Thus, the World Bank works with USAID. In the case of MYWO's MCH/FP program, the World Bank provides financial assistance to MYWO in terms of loans, and defers to USAID for program implementation. Peace Corps volunteers work independently of USAID, but are able to request financial assistance in the form of grants from USAID for certain "training" projects. Pathfinder also implements USAID policies, even though it sometimes resists—albeit unsuccessfully.[16]

Three organizations—GTZ, KAF and Marttaliitto—claimed that they were not beholden to their home governments' foreign policy agendas, but their claims are questionable. The reason is that their funding and their future operations with MYWO in Kenya were, to varying degrees, determined by their home governments. It is important to note, however, that these organizations were not as closely linked to their home governments as the organizations previously discussed.

KAF, for instance, was accountable to the German government for the monies spent for MYWO's LD program. When KAF could not demonstrate to the German government that the monies were put to "good" use during the last program period of 1988–91, the program was not funded for the next period. MYWO had not turned in required accounts records for this period.

GTZ and Marttaliitto also depended on their home government's approval of their continued interorganizational relations with MYWO. The renewal of their operating budgets was based on evaluations of program performance by their home governments.

Marttaliitto appeared to be the most transparent foreign donor with its agenda. The focus of its project was on food, nutrition and environmental technology as a strategy to protect against times of hardship. At the onset of its partnership with MYWO and the Kenyan government, Marttaliitto contributed 60 percent of the funding for the project and the Finnish government provided 40 percent. During the course of the project, the Finnish government increased its contribution to 75 percent and Marttaliitto contributed 25 percent.[17] GTZ also appeared to have no underlying agenda in the partnership with MYWO and the Kenyan government.[18]

This raises some important questions: What constitutes "substantial" contributions from a Northern government to a Northern NGO? At what point is the latter's autonomy compromised? Does the compromise of its autonomy mean it is dependent on the state? Does compromising its autonomy mean that it adopts the policies and agenda of the home government? If that is the case, *can* foreign private donors be transparent? To what extent do Northern governments and Northern foreign donor relationships differ, for certainly Northern governments differ in their ideologies and approaches to development in the South? Are the approaches to and intentions of development different among government agency donors and (substantially) government-funded private donors? Why do governments assist private donors?

It is important to note that some donors, such as Marttaliitto and GTZ, had genuine commitments to grassroots development. As a group, however, the donors' objectives varied widely. Few donors were as transparent as

Marttaliitto. More donors had political and clandestine objectives, such as the promotion of liberal democratic and capitalist ideologies, sometimes at the expense of development (albeit for the short term, they would argue). These donors, including USAID and the World Bank, are the stalwarts and watchdogs of the global interventionist strategy of this decade and have dictated the adoption of structural adjustment economic reforms and multi-party liberal democratic political reforms for the South.[19]

On one hand, it appears that MYWO's national leadership complied with foreign donor program and project implementation, often without substantial input, in exchange for foreign donor assistance and its spillovers—spillovers which enhanced their personal and class interests. On the other hand, it appears that foreign donors cooperated in development partnership endeavors with MYWO, in spite of MYWO's organizational problems, in order to continue the implementation of their projects. Concomitantly, and maybe even more importantly, foreign donors perhaps maintained relations with MYWO specifically to have influence in Kenya's internal political and economic affairs. MYWO (as agents of KANU) maintained relations with foreign donors to have access to their finances. Thus, these partnerships between foreign donors and MYWO (and by extension KANU) were ones of mutual dependence and exploitation between partners. The principal reason MYWO maintained that it was an NGO, despite being taken over by KANU, was because this was its strategy to further exploit foreign donors.

MYWO appeared to have more autonomy from some foreign quasi-governmental donors in program and project implementation—those less directly associated with GOs and IGOs, e.g. Marttaliitto, GTZ and KAF—than it did from the business Coca-Cola–Kenya. MYWO appeared to have least autonomy from foreign and international governmental donors (GOs and IGOs), e.g. USAID and the World Bank, and from quasi-governmental donors more directly associated with GOs and IGOs, e.g. CEDPA.[20]

Overall, foreign donors—governmental, quasi-governmental and business—seemed to have more power than MYWO in development partnerships as they could implement programs and projects more or less as they pleased, and/or they could end partnerships when they chose. For example, USAID had the most power in determining family planning issues relative to all the governmental and quasi-governmental organizations involved because of USAID's bilateral terms and arrangements with the Kenyan government. There was also strong donor solidarity among foreign donors—governmental and quasi-governmental—which they used as a political bargaining chip to flex their muscles and get what they wanted in Kenya and with MYWO. For example, foreign donors banned together in support of NORAD during NORAD's disagreements with the Kenyan government which led to Kenya's eventual severance of diplomatic ties with Norway in 1990. Foreign donors, as a group, refused to take over MYWO projects that had been funded by NORAD. Some of those projects are still not covered. It is important to note that while donors flexed their muscles to show their strength in interorganizational relations, and while the Kenyan government demonstrated its sovereignty, it is people, and especially women at the grassroots, who suffered because they no longer received development-related family planning services.

This research further determined that foreign donors have more power in enforcing the adoption of their policies than both MYWO and the Kenyan government. MYWO's and the Kenyan government's dependence on foreign donors for assistance, coupled with the lack of will of their leaders, make them unlikely to effectively resist foreign donor demands. Although this runs contrary to the arguments of many African state sympathizers from the pessimist school, such as Kinyanjui, Kobia, and Tandon, it is partly by choice that the Kenyan government and MYWO do not effectively resist.[21] As demonstrative of this, the Kenyan government's themes and policy framework of its Sixth Development Plan, for the years 1989–93, fundamentally mirrors the strategies for structural adjustment proposed by the World Bank and the IMF, with apologies by the government for its poor performance to date.[22]

In light of this, it is no wonder that most foreign donors seemed to "bring" their development agendas with them. When it is understood that MYWO implements supplemental programs to that of the Kenyan government, which has embraced World Bank and IMF policies, it is then clear that MYWO has no indigenous grassroots women's agenda, partly because the Kenyan government has no indigenous national agenda and perhaps because MYWO is out of touch with local women. When it is understood that MYWO receives the bulk of its funding for its operations and programs from institutions creating policy priorities for major multilateral institutions and bilateral institutions which heavily influence governments, it is then clear that MYWO *cannot* really have an indigenous agenda. When it is understood that KANU and MYWO leaders are in class positions which allow them to benefit economically and politically from "going along" with foreign donor agendas to get more assistance, it is then clear that MYWO *does not* desire to have an indigenous development agenda. Moreover, it is clear that in KANU's clever maneuvering to capture foreign donor funding through MYWO, it perhaps has captured a global interventionist development agenda that cannot be easily, if at all, modified.

Just as MYWO has no real autonomy from KANU, this research found that MYWO has virtually no real autonomy from foreign donors either, particularly since MYWO is a resource-dependent partner in development interorganizational relations. The fact that 79 percent of MYWO's funding for *all* of its operations comes from USAID calls this issue to question. Can MYWO be autonomous when it receives such an overwhelming amount of its resources from foreign donors? To further complicate this matter, governmental and quasi-governmental foreign donors do not appear to have complete autonomy from their home governments either. Smith, Lethem and Thoolen, and also de Graaf and Twose, have each demonstrated quite correctly that organizations must interact in a larger political and economic environment which extends beyond the immediate environment of development implementation. They argue quite effectively that in that larger environment Northern donors are necessarily connected to their home governments.[23] This research has further shown that the extent to which Northern donors and Northern governments are connected differ from country to country, as well as by the type of government and type of donor. Whether or not this makes a difference to the nature of developing interorganizational relationships and to development outputs and outcomes in the South is a question that must be probed.

This connection between Northern donors and their home governments proved to be a fundamental part of MYWO's interorganizational relations. This research demonstrated that interorganizational relations become hypocritical when foreign donors withdraw funding or make threats thereof because of MYWO's affiliation to KANU. The reason is that these foreign donors themselves are affiliated to their home governments or political parties, and many of them push policy objectives and ideologies on behalf of their governments. It is further hypocritical that these donors make charges that KANU is using MYWO, when many foreign donors are themselves using MYWO. In many instances, it was the fundamental belief that MYWO could, in fact, be used that was the critical motivation for the creation of development cooperative partnerships with MYWO in the first place.

To label foreign donors' criticisms and interorganizational behavior hypocritical is not to suggest that KANU's affiliation of MYWO was justified; nor is it to suggest that KANU should *not* be justly criticized for its exploitation of MYWO. To label foreign donor criticisms and behavior hypocritical is merely to point out that, in the politics of development cooperation of MYWO for instance, there are conflicting standards by which development partners are supposed to abide. On one hand, it is acceptable for Northern donors, both private and governmental, to be associated with their home governments; but on the other hand it is not acceptable for MYWO to be associated with its home government. Moreover, a major undercurrent in development "cooperation" is that Northern donors, on behalf of their governments, have a right to exploit their "partners" and have a responsibility to act on behalf of their governments as well as be exploited by their home governments. Southern NGOs, quasi-NGOs and governments, however, do not have those same rights and responsibilities, and perhaps should feel "morally" opposed to exploitation. It is almost as if we do not expect the North to oppose exploitation, as we legitimize their exploitative actions "on behalf of their governments" or as "necessary for national security." When the South does not oppose exploitation, however, but instead participates in it, we refer to its actions as amoral and corrupt. Stated differently, somehow we *expect* the North to be exploitative and sinister and the South to be non-exploitative and demure.

This double standard framework, while applicable on one level, by itself offers an outmoded over-simplistic framework for understanding IOR and the politics of development cooperation. That is, interorganizational relations between organizations in the North and the South can no longer be analyzed from the perspective that the North's governmental and non-governmental institutions are the calculating villains and the South's governmental and non-governmental institutions are the innocent victims. IOR relationships and networks have become extraordinarily complex and IOR political actors too strategically interdependent and manipulative of their environment to be understood and explained by unsophisticated models. More applicable models must, for instance, consider MYWO elite women's power maximizing strategies in interorganizational relations.[24]

Kenyan government ministry bureaucrats/personnel

The Kenyan ministry bureaucrats represented an often ignored link in development interorganizational relations. They were a very frustrated local Civil Service who had been drawn into this development network reluctantly. As a group they remained significantly delinked from MYWO's interorganizational relations, to a large extent by choice. The female personnel, in particular, despised MYWO. They also remained subjectively delinked from "the government."

The ministry personnel were, for the most part, very knowledgeable, highly skilled development professionals, except for the MCSS who were not technically skilled. They were university graduates, and many prided themselves that they were not politicians. Of all of the groups of interviewees, they had the most realistic sense of grassroots development issues. Ironically, these bureaucrats represented the lowest rung of development consultants. Their commitment to development, for the most part, was surprising and commendable considering their very low rates and undependability of pay. Even more surprising is the fact that most of them did not complain.[25] These findings are important for building knowledge bases of Southern bureaucratic behavior.

Although each ministry had a unique relationship with projects that may have involved MYWO, it was near consensus that MYWO was despised by most of the ministry bureaucrats. MYWO was seen as nothing but KANU. They were very critical and loathsome of MYWO. Their belief was that MYWO had become an inept parasitic organization with which they would have no involvement. They felt that MYWO was a time bomb that would blow up with KANU. They saw MYWO as neither autonomous from

KANU nor indigenous. They regarded MYWO's disaffiliation from KANU as mere symbolism to remain tied to foreign donors.

MYWO was not considered a partner in development by ministries. Partnerships were identified as existing between the foreign donors and the Kenyan government—ministries which were the implementing arm, notably separate and apart from MYWO networks.

Ministry personnel were also critical of foreign donors. They believed that providing foreign technical experts was a way to impose a foreign agenda on Kenya and development groups.

The ministry views corresponded somewhat with the views of the unintended group of interviewees who emerged on the *magendo* as "other relevant interviewees." They believed that historically MYWO had been detached from grassroots women. They argued that MYWO had never really caught on at the grassroots. From the colonial times of European leadership to the time of its disaffiliation from the colonial state and its Africanization, MYWO had been a middle-class women's organization.

This group was not surprised that KANU had affiliated MYWO. The pattern of the government was to control entities that it could not destroy. Bratton has noted that one of governments' reactions to hegemony-threatening organizations is cooptation.[26] For the Kenyan government, it would have been politically and financially unfeasible to blatantly destroy MYWO, as public opinion would have condemned KANU as sexist and non-populist, and foreign donors would have withdrawn all or substantial parts of funds used for Kenya's development. (The latter assuredly carried more weight in the government's decision.) Thus, it was to KANU's benefit for its purse and public opinion to become a member of a development cooperative partnership with foreign donors and MYWO. KANU manipulated MYWO in order to do so, by imposing itself on MYWO. Thus, KANU was seen as a manipulator of the women's organization. Foreign donors were also seen as manipulators of MYWO and the Kenyan government, using their finances to push their foreign policies and ideologies. MYWO leaders, who were KANU, were seen as interorganizational co-conspirators.

This research found that ministry bureaucrats felt that all of the partners in development realized their relative gains within development cooperative arrangements, with foreign donors receiving most and women at the grassroots receiving least. They argued that if MYWO were smarter and diversified its funding it would have more power with foreign donors than it did. They argued that MYWO could have taken some action to increase its relative power vis-à-vis donors, but they were blocked in doing so by constraints in their interorganizational relations with KANU. There was also a possibility that MYWO could design development cooperative partnerships so as to have some degree of autonomy from foreign donors as other Kenyan organizations had done. With KANU constraints, with a self-interested MYWO elected leadership who were not development conscious, and with all partners in development realizing some relative gains within development cooperative arrangements, however, there was no desire on MYWO's part to arrange development cooperative partnerships to this end.

This group argued that foreign donors could not be trusted, as they were neither committed to development in Kenya nor did they provide development assistance without strings attached. The unspoken exchange for foreign assistance was that Kenyan national planners would not engage in national policy formulation that would deviate from World Bank strategies, nor would they challenge the conditions of the donors who set the development agenda globally. In 1991, the foreign donor agenda was no longer subtle as it directly linked economic and political reforms as essential conditions to be met in order to receive aid.

The ministry bureaucrats' cynicism extended to the Kenyan government, just as it did to MYWO. They argued that KANU and the Kenyan government could not be trusted either. Both government leaders and MYWO leaders had significantly contributed to the very conditions which "necessitated" the call for reform conditionalities by foreign donors. They were part and parcel of the graft, the inefficiency, the lack of accountability and transparency, the repression and the gross mismanagement of the Kenyan state and the KANU government. MYWO leaders, KANU leaders and the Kenyan government leaders were the very folks who had benefited and entrenched their class positions from these ills. In light of this, the Civil Service could not be effective in implementing development because its top echelon, who were KANU stalwarts, were amenable to being bought off. Moreover, average Kenyans felt powerless to fight those institutions they saw as the enemy, i.e. the World Bank, the IMF, KANU and its appendages, including MYWO. The forces against them were totally inequitable.

Contributions to the literature

A reconsideration of interorganizational relations resource dependency model (RDM)

Considering these findings, this research demonstrated that the creation of development partnerships between foreign donors—governmental, quasi-governmental, and business—the Kenyan government and MYWO was grounded in very specific reasoning, which included but was not exclusive to resource dependencies. Advocates of the resource dependency model, such as Cook, Mulford, Burt, and Blau, focused primarily on 1) the *objective need for resources* as the basis for the creation of interorganizational relations, and 2) the nature of the interorganizational relationships that resulted from creating unequal resource interdependencies.[27] As this case study of MYWO and the organizations in its relational environment has demonstrated, there were other propelling forces which provided the bases for establishing interorganizational relations. These were *subjective* motivations for the creation of interorganizational relationships which underlie organizations' objective resource needs, and which, over time, shape the character of the interorganizational relationship. For instance, in this case study the primary reasons for establishing interorganizational development partnerships were:

1 MYWO's lack of financial, technical and material resources to implement its programs and projects on its own, at the level that the donors provided;
2 the KANU government's lack of financial, technical and material resources to implement its programs and projects on its own, at the level that the donors provided;
3 foreign donors' lack of organizational access in Kenya which could provide them the necessary clientele to serve as well as consumers and supporters of their programs;
4 MYWO's desire to have financial partnerships and bring more monies into the country;
5 MYWO elected leadership's desire to benefit financially as individuals;
6 KANU leadership's desire to bring more money into the country to benefit financially as individuals;
7 KANU's nonprioritization of development programs with government resources;
8 foreign donor countries' desire to have presence and political influence at the grassroots in Kenya.

Reasons 1, 2 and 3 are supported by the resource dependency model, as they represent the objective quantifiable resource needs of MYWO, foreign donors, and the Kenyan government. Reasons 4 to 8, however, pose challenges for the re-evaluation of the RDM. MYWO and its partners in its relational environment clearly had *subjective*, selfish and sometimes sinister motivations and goals, such as greed, domination, control and the maximization of power, for entering into interorganizational exchanges that the RDM does take into consideration. Less apparent secondary motivations not captured in reasons 4 to 8 are: MYWO elected leadership's desire to 1) provide financial and job security for themselves and select staff; 2) solicit monies for KANU; and 3) give the impression for working within the rubric of "women in development" in Kenya. KANU wanted to 1) capture and control MYWO and foreign donor finances; and 2) at the same time give the impression of integrating women into Kenya's development. Additionally, foreign donors wanted to implement humanitarian, environmental, diplomatic and/or political programs and projects in Kenya—usually with economic and political conditions attached. Thus, not all of their interorganizational exchanges were resource driven and objectively defined. This calls into question the overall explanatory power of interorganizational relations' resource dependency model in the politics of development cooperation.

A reconsideration of NGO's two schools of thought: a view towards an alternative model—the web of deceit

The findings of this study indicate that neither of the two schools of thought on NGO partnerships—the optimist school which ignores power differentials in relationships and the pessimist school which maintains the villains and victims perspective solely—offer accurate and acceptable explanations for development cooperative relationships in this case study.[28] One reason is that the optimist school is too romantic and the pessimist school's Southern victim–Northern villain dichotomy is too simplistic. Another reason is MYWO's schizophrenic organizational personality which prompted its shifts between NGO and GO identities at select times. These shifts facilitated its relations with its interorganizational partners in development on one hand and conflicted those relations on the other hand. MYWO's chameleon identity is a direct consequence of its lack of organizational autonomy.

This research found that what is really at issue in the politics of development cooperation between MYWO, the KANU government and foreign donors is a development game of users grounded in exploitation, expedience, hypocrisy, convenience and corruption, in which MYWO and the Kenyan government and foreign donors in its relational environment all play a part—and the victims and villains are intertwined between the North and the South and between women and men. In this interorganizational network of development partners, there exists a "web of deceit" in which all of the partners are relative strategic winners, and in which those who are the consistent losers are the women at the grassroots. They are the losers because it is on their reputation that the "women in development" ideology is based. It is on their reputation that bogus development cooperative partnerships have been created, as grassroots women are the imaginary direct beneficiaries of development programs implemented by MYWO and its donors in its relational environment. Patricia McFadden, in her work with women's organizations in Southern Africa, insightfully notes that "grassroots" is often not used to mean the mobilization and resistance of poor women and men. Instead, it is often used to reinforce the elitist notion that certain women are at the bottom of the social scale.[29]

This research demonstrated that as KANU uses MYWO to secure monies from foreign donors, a chain reaction occurs—foreign donors use MYWO to implement their programs and projects; MYWO national elected leaders use women at the grassroots to build a development reputation; MYWO elected leaders use KANU and foreign donors for personal gain; foreign donors use MYWO and KANU to implement market-based economic strategies and multiparty political systems; and ultimately the web of deceit becomes more intertwined and enlarged while the presumed goal of "development" is ensnared and forgotten. Generally, this research determined that there was very little interorganizational sincerity between and among most partners in development; and there was no indigenous grassroots agenda to which they addressed themselves. Most partners had a very selfish agenda which had little or nothing to do with development at the grassroots. For the leaders of MYWO and the KANU party, interorganizational relations had to do with personal enrichment and political power, literally at the "expense" of foreign donors. For many foreign donors, interorganizational relations ultimately had to do with maintaining the divisions between the North and South through the implementation of paternalistic, often ill-conceived policies (such as structural adjustment) that are fallaciously called "development." The attitude of suspicion that Hirono spoke to is part and parcel of IOR relations. Mutual hostility is disguised, however, although this disguise was unraveling in MYWO's and KANU's actions towards foreign donors in 1991–2. The issue of mutual insincerity which the literature did not address was most critical and problematic.[30]

To expose this web of deceit is not to suggest that there are not individuals and organizations in both the North and the South who are sincerely committed to development. There are; but as Kobia of the pessimist school suggests, they are mere pawns in a very sinister anti-development game. Those committed to development are the real development practitioners—women at the grassroots, local Kenyan ministry bureaucrats and some smaller foreign donors (for there does exist a hierarchy among donors).[31] These groups are strongly committed to people-centered grassroots development, but they are a much smaller and less powerful force that operates on the periphery of the politics of development cooperation. As their awareness grows, many of them—women, civil servants in the ministries and foreign donors—delink themselves from the major international and national development networks.

Sithembiso Nyoni and Bertrand Schneider separately argue that, instead of delinking, rural people and local governments ought to be actively included and should remain in interorganizational development partnerships with foreign donors. Nyoni, for instance, urges INGOs to link with local groups instead of promoting "international developmentalism"[32] and Schneider admonishes governments to train their civil servants and motivate them to be functional participants in development implementation.[33] How this is to be done, however, is a question still unresolved. This study, for instance, has demonstrated the extent to which people at the grassroots and local governments are actually marginalized and overshadowed in the politics of interorganizational development cooperation.

This research further found that issues of integrity, authenticity and will lie at the very foundation of development cooperative partnerships. For example, at a very fundamental level, donors, both Kenyan government and foreign, did not respect MYWO as an organization. Neither did they consider MYWO a partner in development—not necessarily because it had fewer resources, but rather because it had no principles or integrity. This only exacerbated the fact that foreign donors and the Kenyan government were already less inclined to have respect for MYWO in interorganizational relations, because it is a women's organization.

In this web of deceit, MYWO presented itself as an NGO committed to the development of women and of Kenya, but lacking in the necessary resources to implement development projects and programs. This research found that MYWO is a chameleon with multiple personalities, and the NGO persona was one of its façades. MYWO at the national level was basically a "tea club" for KANU women; it was a political appendage of the KANU party and government. MYWO was not a committed grassroots development NGO for women, although it posed as such. All of the critical political actors in development partnerships knew that MYWO was not an authentic development organization; women at the grassroots certainly knew it. MYWO was, in effect, *not* a women's organization with a women's agenda. Its leaders had undermined its genderedness and patronized women at the grassroots. Fundamentally, MYWO leaders had no will.

KANU had helped to create this veneer for MYWO with the complicity of MYWO elected officials. Some donors, like Marttaliitto, acknowledged this and, despite MYWO's denial, decided to withdraw from their partnership. Other donors, like USAID, knew that MYWO was not a committed development organization but did not acknowledge it because they wanted to continue their development and/or political work. Skeptics might argue that some donors may have been drawn to MYWO purely because it was *not* a committed development women's organization.

KANU, both party and government, also lacked integrity. It has a longstanding history of dishonesty, fraudulence, untrustworthiness and craftiness. As an institution, its reach had extended far beyond democratic governance to autocratic control and repression. This repression peaked in its relations with MYWO and foreign donors during the period of my research. KANU wanted to dominate all political, economic, social, cultural and developmental institutions. It effectively managed none. Moreover, KANU was paranoid about potential opposition, including women's organizations. It was threatened by its belief that women's power, real and imagined, would loosen the pillars of patriarchal control. It was also threatened—perhaps most threatened—by foreign donors and their economic and political power. Like MYWO, KANU party and government leaders did not have the will to implement development in Kenya.

Foreign donors, like MYWO, are also chameleons. Many of them became involved in development partnerships under the guise of desiring to implement development in Kenya by providing financial and technical assistance. It is no surprise that many of them attempted to bypass Kenyan government corruption, mismanagement and red tape by going directly to people's organizations. As a result, however, foreign donors have demonstrated their potential and power to undermine the sovereignty of the Kenyan state.

Furthermore, they are also contributing to the myth that Northern and Southern NGO partnerships are leading development in Africa.

Moreover, many foreign donors have dubious development goals, since most originate in countries (as Tandon argues) whose governments have tainted histories with the South in general and with Africa in particular. As a result, many questions linger with regard to their integrity and will. Cynics might even argue that foreign donors have used MYWO and KANU dependence and greed in interorganizational partnerships in attempts to capture the Kenyan state.

With regard to the issues of autonomy and dependence, both of which suffer from imprecise definitions, the following observations were made. Since 1952, MYWO has not been truly autonomous. Even when not officially affiliated to the government between 1961 and 1987, and then after 1991, MYWO was still tied to the state by its leadership who were wives and relatives of KANU men. Because of gender discrimination, these women were not genuinely accepted members of KANU. Thus, for some women, to be in MYWO leadership meant they were one step closer to their patron, KANU. Dependence was somehow positive and functional for them.

MYWO had little autonomy from most foreign donors in terms of determining policy or funding during the period of my research. When MYWO did have some degree of leverage or autonomy, it was to confer with and defer to KANU for direction, since MYWO was beholden to KANU for "saving it from itself." Thus, MYWO had no indigenous policy or agenda of its own. The irony of it all is that although MYWO was to defer to KANU, Kenyan government policy was really foreign donors' policy. Thus, the overall development policy framework was basically that of the World Bank and the IMF, for the Kenyan government had been placed and had placed itself in a position of adopting the development policies of these IGOs.

This suggests that in many instances it is useless to negotiate "policy" in development "partnerships" between the North and the South, since it is foreign donor policy that the partners will usually embrace anyway, whether they agree with this policy or not. Much of this also depends on the foreign donor's relationship to the global policy-making institutions of the World Bank and IMF, coupled with the will of Southern leaders. Many times, in development interorganizational relations, the lesser partner has little real option to say no. As this research has demonstrated, the lesser partner may not even have the will and the vision to find out what the foreign donor policy and goals are, since this partner is too often blinded by the monetary assistance it will invariably gain.

In MYWO's development cooperative partnerships, foreign donors usually set the stage with policy initiatives and supporting financial, technical and material resources. MYWO was the vessel through which many foreign donors channeled their ideas and their monies. KANU was the cosmetic regulator of the partnership and the real coveter of foreign donors' financial support of MYWO. KANU and foreign donors had little direct communications with matters relating to MYWO. MYWO was the conduit.[34]

MYWO was the ideal choice for this arrangement with foreign donors and KANU, despite its not being a bona fide NGO. Both foreign donors and KANU could capitalize. As women, MYWO was likely to be reliable for implementing their "development" agenda; and as corrupt politicians, MYWO leaders were not likely to be resistant to the demands of their partners. Moreover, women at the grassroots, silenced by fear and limited information, would not challenge MYWO at the national headquarters. MYWO was ideal in this development cooperative partnership because women tend to be an effective buffer between the imperial strong arm of many foreign donors on the one hand, and the easily bruised male ego of the African state on the other.

Over a decade ago Johan Galtung argued, quite insightfully, that international organizations such as the World Bank and its related organization the IMF would be exposed to increasing criticism because of their imperialist roles. He stated,

> International organizations may become giant mechanisms through which people in the stronger states that started these organizations can imprint a message on the people of the weaker states: "You must have this ministry and that profession, this hobby and sports association and that ideological movement, you must produce this and that in order to be full-fledged members of the World."[35]

The extent to which the Kenyan state and MYWO elite women would support this move by international organizations as their co-conspirators was underestimated.

PART II: CROSS-NATIONAL COMPARISONS, RECOMMENDATIONS AND CONCLUSIONS

Prospects for comparisons

De Graaf has provided some observations of interorganizational relations in Zimbabwe which offer a springboard to beginning systematic comparisons cross-nationally. He has made three observations which are similar to the patterns which I have observed in this study. De Graaf has noted: 1) both weak and strong linkages between NGOs and government; 2) NGOs look more to foreign donors rather than to government for assistance, and government is seen more as a higher political authority than as a resource donor; 3) women's NGOs tend not to have particularly feminist consciousnesses, and do not necessarily push for changes that positively affect the condition of women.

Whether or not these patterns will be replicated in other countries will be determined as other studies involving IOR, NGOs, quasi NGOs, GOs and women are conducted. A forthcoming series of studies on human cooperation and global networking promises to bring further insight into this matter.[36] From its contributions, along with those of de Graaf and those of this study, we may then consider whether or not this cumulative knowledge will be able to generate some lawlike generalizations or theories about IOR, NGOs, quasi NGOs, GOs and women in the partnerships and politics of development cooperation.

Recommendations for practitioners and scholars

For those in the field in Kenya who are implementing development programs and projects, the web of deceit is woven so tightly that one almost feels defeated before tackling it. For the grassroots to experience and implement development, however, that is

> people's participation without fear in bettering their life conditions and prospects, in an enabling environment free from structural violence of institutions and ideologies in both their immediate and extended relational environment.[37]

the web must be tackled. Recognizing that the web exists is the first step. Delusions that development as currently implemented through MYWO in Kenya is anything but a web of deceit must be discarded, if the development game is ever to end.[38] Women's groups' reputations (real and conjured) should no longer be exploited by national organizations such as MYWO, political parties, African governments or foreign donors. Women need bona fide women's organizations if they are to continue to work in women's groups and towards development. MYWO is but a sham and new elections are not likely to produce significant changes.

Attitudes and practices of women, MYWO leaders, government leaders and foreign donors must also change. Women at the grassroots must realize and hone the potential that they have, for they contribute tremendously to development. Women at the grassroots, MYWO staff and elected officials must release their beggars' frame of mind, and relinquish this psychological dependency on "money from the outside." As Nzomo argues, women must also rise above passivity towards men and domestic and foreign institutions. Moreover, women must stop using "culture" and "tradition" to justify their oppression.

The attitudes and practices of MYWO and KANU elites must also change. We must be prepared for more deceit, however, for when we ask them to acknowledge the development game and end it, we are asking them to reprove their *wabenzi* and comprador class positions. They are not likely to change, since many of them already realize their anti-development ways.

They are cognizant that they are part and parcel of this game. They are the linkage which allows the web of deceit to remain firmly entrenched.

With regard to foreign donors, many of them maintain attitudes and actions which must change if development in Africa is ever to occur while they are present on the continent. The indirect forces that support those foreign donors who may have paternalistic and/or clandestine motives must also change. International relations between the North and South are structured so that, in order for a foreign donor to effectively change its behavior and goals in interorganizational relations, the fundamental relations between nations must change as well. For relations between nations to change, world systems and world values must first change. Basically, the entire environmental context within which foreign donors and organizations exist and function must change. The reason is that organizations are not mere products of themselves; they are products of a larger interconnected political and economic environment. In addition, development must be left to the people, for them to guide and for them to be *assisted* by trained development professionals, not politicians.

This study attempted to bridge literature gaps between four seemingly disparate bodies of literature, and future research should additionally synthesize this literature and close this gap. Gender should be central to further investigations. More research should be done on women's participation in development, raising questions about the legitimacy, indigenousness and preponderance of women's organizations. Nzomo, for example, demonstrates that only a small fraction of women in Kenya—1 in 11—belonged to women's groups at the close of the 1980s.[39] Women in Development (WID), Women and Development (WAD) and Gender and Development (GAD) ideologies and their translations in program implementation and results for women should also be more closely scrutinized.

Regarding NGOs, there is much work to be done for understanding this phenomenon, in and of itself and in partnership with other institutions, both in the North and in the South. NGOs remain, to a large extent, anomalies to development practitioners and scholars. They have yet to be clearly defined and analyzed as institutions involved in development partnerships. There are probably exponentially more quasi-NGOs like the ones examined in this study.

Research should also focus more on the interdependence of world systems—governmental, non-governmental and quasi-governmental. The world's move towards globalism in the twenty-first century necessitates this. Whether or not functional autonomy of women's organizations or other organizations can exist in this interdependence is a critical question to be investigated. Esman and Uphoff have already argued that functional dependence of NGOs on states is important for development implementation in the South.[40]

Lastly, scholars must reconsider the definitions of "development success" and "development failure." This research found that determining project success or project failure was not given high priority in the field by the partners in development. Although academicians might think that determining the success or failure of a program or project is important (and it is), this is not what the real development cooperation "game" on the ground is about. In this case study, development on the ground among partners was *more* about foreign donors and national political elites, men and women, satiating their political and personal appetites, than it was about meeting needs and enabling populations at the grassroots. For scholars, another reconsideration of what is meant by "development" is imminent.

In conclusion, this study suggests to development practitioners and scholars that Africa's development crises may not be solved by development cooperative partnerships. Development optimists may be romanticizing the positive prospects that development partnerships are projected to bring to the South. Development pessimists are not hopeful for Africa's development through development cooperative partnerships, and this study of MYWO and its partners in development provide more support for their skepticism. While many development pessimists put the blame for lack of development of the South on the North, they fail to point to the culpability of some partners in the South with the North. These culprits must be recognized as part and parcel of the peril of Africa's development crisis if the continent is ever to progress.

It was clear from the study that women from the South are used by a multiplicity of sources in a very dangerous global patriarchal development game. Some of the users are men (and women) from the North. Some of the users are men from the South; and some of the users are themselves women from the South. All of the above participants are players in this patriarchal game of money, power and politics. Feminist theory and development theory must reconcile the fact that class and political considerations by women often override their gender solidarity considerations. Both Staudt and Nzomo each have implied this in their work. As such, WID, WAD and GAD ideologies must be re-evaluated. We must also ponder, as Parpart suggests, the degree to which Northern paternalism is built into these ideologies.

This study further illustrates that there are motivating factors behind the establishment of interorganizational relations, other than resource necessities, which IOR's resource dependency model does not capture. Greed, domination, control and power are those critical factors. They proved to be the axes on which development interorganizational relations turned.

The work also suggested that autonomy of Southern NGOs may not be as important a consideration as functional dependence, if that dependence brings organizational elites personal satisfaction. That is, autonomy of a Southern NGO may not be a desirable organizational feature in development cooperative partnerships to partnership leaders, since they may not have the will and commitment to development and to people at the grassroots in the first place.

The (im)balance of power in interorganizational relations is firmly entrenched, reflecting power differentials in international relations between states. Development cooperative partnerships may not be interested in challenging these power differentials. Instead, they are mere microcosms of these differentials. In light of this, real development cooperation cannot exist if there are no challenges to the international power structure. Women from the South are not likely to significantly change this structure.

If the web of deceit, created by partners in the politics of development cooperation as well as by forces in their relational environment, continues to spin and entangles more of the continent, we, the global community, must ask ourselves: What hope does Africa have for development today and self-reliance in the new millennium?

Caught in this web of deceit, as victims, will remain some of the most exploited—the grassroots, particularly women. Their struggle to disentangle themselves will not be easy, and the reasons for that are clear: 1) African political leadership in development cooperative partnerships, prodded by the demands of global politics, economics and patriarchy, is likely to continue to invite the exploitation of the continent; and 2) international relations and interorganizational relations in development cooperative partnerships are not likely to be sufficiently restructured to bring about non-exploitive and appropriate assistance to the grassroots.

That a few grassroots women's groups manage to successfully elude the web of deceit is evidence that some of us, however few, do not have to be "caught."

APPENDIX A.I

HYPOTHESES

The hypotheses for this study were derived from the current state of knowledge on 1) the schools of NGO optimism and NGO pessimism, 2) IOR theory, and 3) MYWO. The four major hypotheses that were proposed in this study are:

1 The greater the financial and technical assistance MYWO receives from foreign donors and the Kenyan government, the less autonomy MYWO will have to formulate and implement its indigenous agenda, hence the less successful MYWO development projects and programs will be.
2 The less the financial and technical assistance MYWO receives from foreign donors and the Kenyan government, the more autonomy MYWO will have to formulate and implement its indigenous agenda, hence the more successful MYWO development projects and programs will be.
3 The more directive the assistance MYWO receives from foreign donors and the Kenyan government, the less autonomy MYWO will have, hence the less successful MYWO development projects and programs will be.
4 The more non-directive the assistance MYWO receives from foreign donors and the Kenyan government, the more autonomy MYWO will have, hence the more successful MYWO development projects and programs will be.

APPENDIX A.II

SUB-HYPOTHESES

The following sub-hypotheses branch from the four major hypotheses that were proposed for testing in this research:

1 (a) The greater the financial assistance MYWO receives from foreign donors, the less autonomy MYWO will have to formulate and implement its indigenous agenda, hence the less successful MYWO development projects and programs will be.
(b) The greater the technical assistance MYWO receives from foreign donors, the less autonomy MYWO will have to formulate and implement its indigenous agenda, hence the less successful MYWO development projects and programs will be.
(c) The greater the financial assistance MYWO receives from the Kenyan government, the less autonomy MYWO will have to formulate and implement its indigenous agenda, hence the less successful MYWO development projects and programs will be.
(d) The greater the technical assistance MYWO receives from the Kenyan government, the less autonomy MYWO will have to formulate and implement its indigenous agenda, hence the less successful MYWO development projects and programs will be.

2 (a) The less the financial assistance MYWO receives from foreign donors, the more autonomy MYWO will have to formulate and implement its indigenous agenda, hence the more successful MYWO development projects and programs will be.
(b) The less the technical assistance MYWO receives from foreign donors, the more autonomy MYWO will have to formulate and implement its indigenous agenda, hence the more successful MYWO development projects and programs will be.
(c) The less the financial assistance MYWO receives from the Kenyan government, the more autonomy MYWO will have to formulate and implement its indigenous agenda, hence the more successful MYWO development projects and programs will be.
(d) The less the financial assistance MYWO receives from the Kenyan

government, the more autonomy MYWO will have to formulate and implement its indigenous agenda, hence the more successful MYWO development projects and programs will be.

3 (a) The more directive the assistance MYWO receives from foreign donors, the less autonomy MYWO will have, hence the less successful MYWO development projects and programs will be.

(b) The more directive the assistance MYWO receives from the Kenyan government, the less autonomy MYWO will have, hence the less successful MYWO development projects and programs will be.

4 (a) The more non-directive the assistance MYWO is able to receive from foreign donors, the more autonomy MYWO will have, hence the more successful MYWO development projects and programs will be.

(b) The more non-directive the assistance MYWO is able to receive from the Kenyan government, the more autonomy MYWO will have, hence the more successful MYWO development projects and programs will be.

APPENDIX A.III

KEY VARIABLES AND THEIR DEFINITIONS

Figure A.1 illustrates the graphic interplay of the variables which will be discussed in this section.

The independent variables (IVs) for this study are:

1 the amount of **financial assistance** provided by foreign donors for MYWO's programs and projects;
2 the degree of **technical assistance** provided by foreign donors for MYWO's programs and projects;
3 the amount of **financial assistance** provided by the Kenyan government for MYWO's programs and projects;
4 the degree of **technical assistance** provided by the Kenyan government for MYWO's programs and projects;
5 the extent of **directive or nondirective assistance** that is provided to MYWO by foreign donors; and,

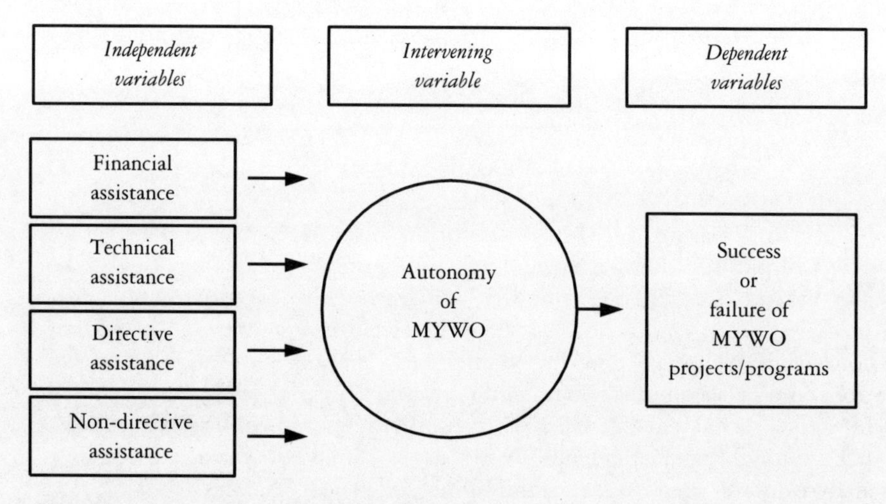

Figure A.1 Presentation of hypotheses proposed

6 the extent of the **directive or non-directive assistance** that is provided to MYWO by the Kenyan government.

Financial assistance is defined as the granting of monies in the form of gifts or loans, 1) from foreign sources outside Kenya which shall be referred to as foreign donors, and 2) from the Kenyan government. This assistance is provided to MYWO for the purpose of assisting MYWO with the implementation of a specified development project, or with a general development program.

Technical assistance is defined as the provision of skills, knowledge, and techniques 1) by foreign personnel who come from outside Kenya who shall be referred to as foreign donors; and 2) by the Kenyan government. This assistance is provided to MYWO for the purpose of assisting MYWO with the implementation of a specified development project, or with a general development program.

Directive assistance is defined as that assistance MYWO receives from foreign donors and/or the Kenyan government which simultaneously brings requests/demands for MYWO development agenda changes made by (a) foreign donors or (b) the Kenyan government which may include, at the extremes, threats of withdrawal and actual withdrawals from organizational cooperative partnerships by (a) foreign donors and (b) the government. **Non-directive assistance** is the diametric opposite of this definition.

These IVs will be considered the instruments of foreign donor and government **influence** in MYWO's relational environment.

Foreign donors providing the types of assistance mentioned above were initially thought to be all foreign NGOs, primarily because there is a tendency in the literature and in lay discussions to refer to all foreign donors as NGOs. Upon closer inspection, however, it was clear that some of the donors were technically not NGOs. Instead they were IGOs and businesses or TNCs, such as the World Bank and Coca-Cola respectively. This called for a reconceptualization of this issue and a recategorization of the external organizations providing assistance to MYWO, from the narrower category of NGOs to the broader category of foreign donors, which encompasses NGOs and IGOs—as well as INGOs, and businesses.

The intervening variable is **autonomy** of MYWO in development program/project formulation and implementation. Autonomy was defined and measured as MYWO's ability and freedom to formulate and implement its own indigenously defined development agenda as initially conceived and agreed upon by its members, making no changes, modifications and/or accommodations to the government and/or to foreign donors in its relational environment for the specific purpose of securing and/or maintaining financial, technical and other resources. Various methods of analyses were used to determine whether or not changes, modifications and/or accommodations in the MYWO development agenda are related to the instruments of influence of foreign donors and the government and, if they are, to what degree.

The dependent variables (DV) are **success** or **failure** of MYWO programs. Success is defined as the state of reaching the desired objectives and goals of development projects and programs as initially defined by MYWO, before these goals and objectives are influenced by MYWO's development partners. Failure is the opposite of success.

APPENDIX B

SURVEY INSTRUMENT: LIST OF OPEN-ENDED QUESTIONS

LOCAL WOMEN AT THE GRASSROOTS

1. How did this project come into existence?
2. Did you (the local women) request the project currently being implemented? From whom? In what setting? If not you, then who?
3. If you requested the project, what role did you play in the formulation of the project?
4. Do you agree with the way in which this project is being implemented? Why? Why not? What would you change?
5. Who do you credit for the presence of this project in this area? its successes? its failures? Is it MYWO, foreign donor or the government? Why?
6. Who takes the leadership role in the implementation of the project? Who makes the decisions, the changes? Should they be the leaders?
7. Who are the assistants or partners in the implementation of this project? What role do they play? Would you change any of the partners in development or their roles? Why? Why not?
8. What power do the partners in development yield relative to each other? Do some partners have more power than others? Are they equals? Why?
9. Do you think that this project is a success or a failure? Why? Why not?
10. What defines project/program success? failure?
11. What are your overall aims in the implementation of this project? Have they/are they being accomplished? Why? Why not?
12. Does your participation in this project affect your power as a woman? How?
13. Who do you credit for the overall development of Kenya? Why?

MYWO OFFICIALS AND STAFF

1 What is your role in development as an NGO?
2 How do you choose which development projects to implement? Who is involved in the formulation of these projects? In what way?
3 What is your relationship with foreign donors and the government? How are relationships with foreign donors and the government created? Are they partnerships? How would you describe them?
4 Who decides which foreign donors will be involved and in what capacity? Who decides how much financial and technical assistance foreign donors and the government will provide? Are there tradeoffs for this assistance? If so, what are they?
5 Are development projects affected when relationships with foreign donors and government are formed? In what ways? Are there changes from the indigenous agenda? Who approves these changes? Why are the changes made?
6 Are there consequences for (a) MYWO, and (b) development prospects when agenda changes are made? What are they? Are there consequences for not making agenda changes? What are they?
7 Do foreign donors and government make demands or requests as partners in development? If so, describe them. What are their reasons for demands/requests?
8 Are there areas of negotiables and non-negotiables between MYWO and foreign donors and governments before and after demands are made? What are they? Why do they exist? Has your organization ever withdrawn from a development partnership? What were the circumstances? Have the government or foreign donors ever withdrawn from a development partnership? What were the circumstances and the impacts?
9 How would you describe the arrangement between the partners in development? Why? Is it an equal partnership or does a hierarchy of institutions exist? Which is more conducive to development? Does any one partner take the lead in development? Which one? Why? Should one? Which one?
10 How autonomous is MYWO? Why? Why not? Does this have an impact on development relationships and prospects?
11 What percentage of MYWO's projects are successes? failures? Are the specific Nutrition, MCH/FP, Energy and Environment, Leadership Development projects successes or failures? Why?
12 What defines project/program success? failure?
13 To which institution or combination of institutions do you attribute project success? project failure? the overall development of Kenya?
14 Do development successes occur because of MYWO partners, or in spite of them? Why?
15 What suggestions would you make for more effective approaches to

development with regard to the NGO, foreign donor, government relationship?

16 In what way does involvement with MYWO affect the power of women?

17 How has MYWO impacted gender equality in Kenya?

18 Does MYWO have a feminist agenda?

19 What are your feelings towards the NGO Coordination Act? the NGO Standing Committee? Is MYWO part of the Standing Committee? Why/why not?

20 What are your feelings towards MYWO's affiliation with KANU? What has been the impact of this affiliation?

21 What do you see for the future of MYWO?

22 What does MYWO need for its continued success?

FOREIGN DONORS

1 What is your role in development as a foreign donor?

2 Why are you involved in the development of Kenya?

3 What are your organization's goals? What are the objectives in the projects and programs you assist?

4 How and when do you communicate your goals and objectives to MYWO? Do you communicate them into the implementation phase of development projects? What happens when your goals differ from those of MYWO?

5 Are there areas of negotiables and non-negotiables when you forge a relationship with MYWO?

6 How would you classify MYWO? Is it an autonomous body with an indigenous agenda? Why? Why not? What impact does this have on your partnership and your financial and technical assistance?

7 How would you describe the arrangement between the partners in development? Why? Is it an equal partnership or does a hierarchy of institutions exist? Which is more conducive to development? Does any one partner take the lead in development? Which one? Why? Should one? Which one?

8 What percentage of the MYWO projects you have assisted in funding have been successes? failures? Are the specific projects which I am studying successes or failures? Why?

9 What defines project/program success? failure?

10 To which institution or combination of institutions do you attribute project success? project failure? the overall development of Kenya?

11 Do development successes occur because of (a) foreign donor presence, (b) government presence? in spite of them? Why?

12 What suggestions would you make for more effective approaches to

development with regard to the NGO, foreign donor, government relationship?

13 What effect does foreign donor participation with MYWO have on gender equality in Kenya? Why?

GOVERNMENT MINISTRY BUREAUCRATS/PERSONNEL

1 What is the government's role in development?
2 What is the government's relationship to (a) MYWO, and (b) foreign donors?
3 Does the government perceive MYWO's development goals as complementary to its own? What happens when those goals are not seen as such?
4 Is MYWO viewed as an autonomous body? Why? Why not? What does this mean in terms of their taking independent development initiatives? Does this have an impact on the financial and technical assistance the government provides?
5 Under what circumstances does the government enter into development relationships with MYWO and foreign donors? Are there areas of negotiables and non-negotiables when the government enters into a development relationship? Has the government ever withdrawn from a development partnership? Under what circumstances?
6 How would the government describe the arrangement between the partners in development? Why? Is it an equal partnership or does a hierarchy of institutions exist? Which is more conducive to development? Does any one partner take the lead in development? Which one? Why? Should one? Which one?
7 What percentage of MYWO's projects which the government assisted in funding were successes? failures? Are the specific projects which I am studying successes or failures? Why?
8 What defines project/program success? failure?
9 To which institution or combination of institutions does the government attribute project success? project failure? the overall development of Kenya?
10 Do development successes occur because of (a) the government or (b) foreign donors? or in spite of them? Why?
11 What suggestions would you make for more effective approaches to development with regard to the NGO, foreign donor, government relationship?
12 What effect does the government's relationship with MYWO have on gender equality? Why?

APPENDIX C

FOREIGN DONORS

UNITED STATES AGENCY FOR INTERNATIONAL DEVELOPMENT (USAID)

USAID is a governmental agency of the United States International Development Cooperation Agency. Its function is to provide economic and foreign assistance through grants to foreign governments and private educational and interest groups for research, employment and training programs, as well as materials for those programs, in the developing world. Since the passing of the 1961 Foreign Assistance Act, USAID programs have focused on: health care, population reduction, education, environment, agriculture, energy, economic growth, human resource development, housing, political freedom and good governance. In 1973, special attention called for USAID policies to address the needs of women in foreign countries, leading to the creation of the WID component of USAID.

USAID in Kenya is an AID Mission because its economic assistance is considered major. The Mission is subject to the direction and guidance of the US Ambassador to Kenya.

USAID is the principal funder for the Maternal Child Health and Family Planning program (MCH/FP) of MYWO National Office via Pathfinder International. This program provides health care services for mothers and children, while focusing on family planning as its primary program. USAID supports other MYWO operations as well.

USAID cooperates with Pathfinder International, World Bank, CEDPA, the National Council on Population and Development (NCPD) of the Kenyan Government, and MYWO as the focal organization to implement the MCH/FP program.

WORLD BANK

The World Bank is a specialized agency of the United Nations which is intergovernmental and international. It states that its aim is to raise the standard of living in the developing countries by conveying financial resources from the developed world through loans. In this case, the World Bank assists in providing financial loan assistance for the implementation of the MCH/FP program. All of its resources for lending come from member industrialized countries. As one of the contributors to the MYWO MCH/FP program, the World Bank's partners in the program are USAID, Pathfinder, CEDPA and NCPD Ministry of Health; MYWO is the focal organization.

UNITED STATES PEACE CORPS

The United States Peace Corps was established by the Peace Corps Act of 1961 and was deemed an independent agency by Title VI of the International Security and Development Cooperation Act. Peace Corps volunteers do, however, act as representatives of the United States Government abroad. They are also subject to the governance of the United States.

The Peace Corps was established to promote world peace and friendship, to assist in training human resource power in foreign countries, and to foster understanding between different peoples of the world. Basically, volunteers aim to assist in the local and national development of their host countries.

Peace Corps linkage to MYWO was through various projects in which volunteers worked, such as small business development and water engineering. Peace Corps volunteers did not work as direct attachés to MYWO. Work with MYWO groups was purely coincidental. Their hosts or partners in development were the Kenyan government ministries, mainly the Ministry of Culture and Social Services. The ministries placed volunteers at their work sites. Peace Corps volunteers had varying but generally very limited interaction with MYWO and their foreign donors.

PATHFINDER INTERNATIONAL

Pathfinder International began as a private fund, initially created in 1957 by Dr Clarence Gamble, a philanthropist. Its aim is to increase the number of individuals in developing countries who have access to family planning.

Through the MYWO MCH/FP program, Pathfinder provides family planning services, teaches technical training and supplies family planning commodities. According to its Kenyan representatives, Pathfinder currently receives 93 percent of its funding as grants from USAID. It considers itself, and is considered by the donor community, an NGO.

CENTER FOR POPULATION AND DEVELOPMENT ACTIVITIES (CEDPA)

CEDPA is based in Washington, DC. It defines itself as a private organization whose objective is to improve the management skills of family planning and health officials of developing nations. Since its incorporation in 1975, it has relied heavily, though not solely, on USAID funds. CEDPA is one of the donors of MYWO's MCH/FP program.

GERMAN AGENCY FOR TECHNICAL COOPERATION (GTZ)

GTZ defines itself as a private organization. It works on some of its projects with the Federal Ministry for Economic Cooperation of Germany. GTZ provides the funding for the Women and Energy project which was launched in 1983 as part of the Special Energy program of the Kenya Ministry of Energy, sponsored by GTZ–German Technical Cooperation. This project focuses on improving the living and working conditions of Kenya's rural areas by reducing fuel-wood requirements in general. In particular, this project focuses on the production and distribution of energy-saving stoves (the Maendeleo jiko) .

GTZ lists as its most active partners: Ministry of Energy, MYWO, Ministry of Agriculture, Ministry of Culture and Social Services, various church organizations and women's groups. As the foreign donor for the Jiko project, GTZ allocates to MYWO a percentage of the budget for jiko activity.

KONRAD ADENAUER FOUNDATION (KAF)

This Foundation is considered a private non-governmental organization, yet its programs are financed out of Germany's public funds. The Konrad Adenauer Foundation states that it aims to promote political education, and practices international cooperation through exchanges of information and contacts, as well as through the provision of aid on a partnership basis. In the developing world, the Adenauer Foundation cites as its goals:

1 liberal democratic political reform;
2 economic liberalization towards the adoption of market strategies;
3 advancement towards self-help;
4 cultural and religious orientation of individual and group development.

The Konrad Adenauer Foundation funds MYWO's Leadership Development program. MYWO and the Konrad Adenauer Foundation are the partners in this program.

MARTTALIITTO

Marttaliitto of Finland is a women's non-governmental organization based in Finland. As a women's home economics extension organization, it became involved in the development of Kenya in 1981. Its stated aim is the development of responsible and enlightened populations in home economics so that families and communities are able to withstand hardships. Marttaliitto does this through teaching nutrition, diet, gardening and food preservation.

Marttaliitto supported the Nutrition program of MYWO. It has also coordinated with the Ministry of Agriculture and Home Extension Offices, though informally. In 1984, the Finnish government decided that it would supplement Finnish NGOs, contributing 40 percent while NGOs contributed 60 percent. In the late 1980s, the government increased its share to 75 percent to the NGOs' 25 percent, with the stipulation that there would now be conditions the Finnish NGOs must meet.

COCA-COLA

Coca-Cola is a transnational corporation which manufactures, markets and distributes carbonated soft drink concentrates and syrups. Coca-Cola soft drink products have been sold in the United States since 1886, and currently they are sold in more than 195 countries in the world. Kenya is one of them. Coca-Cola also manufactures, produces, markets and distributes juice products.

Coca-Cola–Kenya is an independent business in Kenya. It bottles and sells Coca-Cola soft drink products. Coca-Coca Enterprises Inc., with headquarters in Atlanta, Georgia, maintains that the Coca-Cola–Kenya operation is separate from Coca-Cola Enterprises. Kenya bottlers are local businessmen, and over 95 percent of the employees are reportedly Kenyan citizens.

Coca-Cola–Kenya is one of the partners of the MYWO kiosk project, along with MYWO, Nairobi City Commission and the Trade Bank.

APPENDIX D

KENYAN GOVERNMENT MINISTRIES

The following ministry descriptions are taken from a history of KANU documented in *30 Great Years of KANU Handbook 1960–1990* written by the KANU National Secretariat.

MINISTRY OF HEALTH

The Ministry of Health is charged with the responsibility of providing quality health care for all *wananchi* (citizens) so that they may lead economically and socially productive lives. The major health care policies are: 1) increased coverage and accessibility of health services in rural areas; 2) consolidated urban, rural, curative and preventative services; and 3) increased emphasis on maternal/child health and family planning services. As this paper is concerned with MYWO's Maternal/Child Health and Family Planning program, the government then stipulated the following: the 1992 target was to reduce population growth from 3.5 to 3.0 percent and infant mortality from 87 per 1000 live births to 77.3 per 1000 live births.

MINISTRY OF AGRICULTURE

This ministry's objectives include increasing food production, growth in agricultural employment, expansion of agricultural exports, resources conservation and poverty alleviation. In its current operations, the broad areas assigned to the ministry include crop production, protection and marketing, land preparation (irrigation and drainage), soil and water conservation, agricultural inputs and agricultural education and extension. Most closely related to this topic of research is the ministry's initiative aimed at assisting small-scale farmers to boost production of food crops and adopt methods/techniques for protecting crops against insect pests, birds and other environmental contaminants. Home economics extension offices are charged with this responsibility.

This ministry worked with MYWO's Nutrition and Jiko projects.

MINISTRY OF LIVESTOCK DEVELOPMENT

This ministry is responsible for: 1) the implementation of an efficient animal production and marketing system; 2) the provision of appropriate livestock technology for dairy, beef, white meat, honey, etc.; 3) the provision of suitable land and water resources for animal production; and 4) the development of supportive institutional infrastructure for pastoralists. This ministry works closely with the Ministry of Agriculture, and thus indirectly with MYWO.

MINISTRY OF CULTURE AND SOCIAL SERVICES

This ministry has 7 departments. It lists as its function: 1) social welfare and community, 2) community development, 3) national library services, 4) vocational rehabilitation, 5) the Women's Bureau, 6) youth centers, 7) adult education, and 8) sports. MCSS states that it has created programs for the enhancement of the quality of life of Kenyans after independence. For example, in 1976 the government created the Women's Bureau to coordinate women's development activities in collaboration with various women's organizations, including MYWO.

MINISTRY OF HOME AFFAIRS AND NATIONAL HERITAGE

This ministry is charged with the implementation of policies on: prisons, the NCPD, children, probation services, archives, refugees, museums, lotteries, betting and gaming control. Various departments within this ministry are charged with the implementation of specific policies.

This ministry worked with MYWO's MCH/FP via NCPD. It also loaned one of its economists to MYWO for program/project evaluation purposes.

NOTES

CHAPTER 1 INTRODUCTION

1 This definition builds on the classic definition of politics from Harold Lasswell (1936) *Politics: Who Gets What, When, How*, New York: McGraw-Hill.
2 Claude Ake (1996) *Democracy and Development in Africa*, 1. Washington DC: The Brookings Institution.
3 For further discussions on approaches to development, see Michael Cowen and Robert Shenton, "The invention of development" *Power of Development*, Jonathan Crush (ed.) (1995) New York: Routledge, 27–43, and Jane Parpart, "Post-modernism, gender and development" in the same work, 253–65.
4 Walter Rodney (1971) *How Europe Underdeveloped Africa*, Washington DC: Howard University Press; Colin Leys (1974) *Underdevelopment in Kenya*, Los Angeles: University of California Press; A.G. Frank (1969) *Capitalism and Underdevelopment*, New York: Monthly Review Press; Fernando Cardoso and Enzo Faletto (1979) *Dependency and Development in Latin America*, Los Angeles: University of California Press; Peter Evans (1979) *Dependent Development*, New Jersey: Princeton University Press; David Himbara (1994) *Kenyan Capitalists, the State and Development*, Boulder: Lynne Rienner Publishers; Christian Potholm (1979) "The process of African politics" *The Theory and Practice of African Politics*, 139–244, New Jersey: Prentice Hall; Naomi Chazan *et al.* (1992) *Politics and Society in Contemporary Africa*, Boulder: Lynne Rienner Publishers; Rene Dumont (1966) *False Start in Africa*, London: Sphere; Richard Sandbrook (1993) *The Politics of Africa's Economic Recovery*, Cambridge: Cambridge University Press; "Part II: The consolidation of power: the fall of the monarchy, the rise of absolutism, and the definition of the political arena" Irving Markovits (ed.) (1987) *Studies in Power and Class in Africa*, 131–88, New York: Oxford University Press; Ali A. Mazrui (1986) *The Africans: A Triple Heritage*, chapter 11, London: BBC Publications; *Adjustment in Africa: A World Band Policy Research Report*), 1–41, Washington DC: Oxford University Press (1994; Augustin Kwasi Fosu (1992) "Political instability and economic growth: evidence from sub-Saharan Africa" *Economic Development and Cultural Change*, 40 (July), 829–41; Claude Ake (1996) *Democracy and Development in Africa*, Washington DC: The Brookings Institution.
5 John Field (ed.) (1993) *The Challenge of Famine*, Connecticut: Kumarian Press.
6 Walden Bello (1994) *Dark Victory: The United States, Structural Adjustment and Global Poverty*, California: Pluto Press with Food First and Transnational Institute; Jennifer Seymour Whitaker (1988) *How Can Africa Survive?*, New York: Council on Foreign Relations Press; Stephen Hellinger, Douglas

Hellinger and Fred O'Regan (eds) (1988) *Aid For Just Development*, Boulder: Lynne Rienner Publishers; Adebayo Adedeji, "The case for remaking Africa" Douglas Rimmer (ed.) (1993) *Action in Africa*, 43–57, New Hampshire: Heinemann.

7 Much of this data comes from the following sources: John Darton, "Survival test: can Africa rebound?" special series of articles in *The New York Times* 19, 20, 21 June 1994; *Adjustment in Africa. A World Bank Policy Research Report*, Washington DC: Oxford University Press (1994); *The United Nations Development Program 1994 Human Development Report*, New York: United Nations (1994); Daphne Topouzis (1990) "The feminization of poverty" in *Africa Report*, July/August, 60–3.

8 Maria Nzomo, "Beyond structural adjustment programs: democracy, gender, equity, and development in Africa, with special reference to Kenya" Julius Nyang'oro and Timothy Shaw (eds) (1992) *Beyond Structural Adjustment in Africa: The Political Economy of Sustainable and Democratic Development*, 99–117, New York: Praeger.

9 Pierre Pradervand (1988) "Self-reliance in Africa: peasant groups key to continent's survival" *Christian Science Monitor*, 9 March 1988, section 2, 15–16; Pierre Pradervand (1989) *Listening to Africa: Developing Africa From the Grassroots*, 17–18, New York: Praeger.

10 Dr Ruth Nasimiyu, currently a faculty member at the Institute of African Studies at the University of Nairobi, stimulated my interest in this study when, in 1988 at an African Studies Association meeting, she discussed her research on MYWO. She was then a graduate student. Some questions from the audience queried whether or not MYWO could be considered an exemplary women's organization. Other African scholars with whom I have discussed my work have suggested that MYWO may be the largest, as well as oldest, organization of its type on the continent to date. I intend to determine whether or not that is the case in further research.

11 Audrey Wipper's works on MYWO include "The Maendeleo Ya Wanawake movement in the colonial period: the Canadian connection, Mau Mau, embroidery and agriculture" *Rural Africana*, 29 (Winter 1975–6), 195–214; "The Maendeleo Ya Wanawake organization: the co-option of leadership" *African Studies Review*, 18 (1975), 99–120; "African women, fashion, and scapegoating" in *Canadian Journal of African Studies*, 6 (1972), 329–49; "Equal rights for women in Kenya?" *Journal of Modern African Studies*, 9, (1971), 429–42. The first two sources deal more specifically with questions of autonomy.

12 I call these two schools of thought the "school of NGO optimism" and the "school of NGO pessimism." Both are discussed at length in Chapter 2 with other relevant literature.

13 Much of the literature that has been produced thus far is grounded in queries about participatory development, NGOs and state interaction, and state, civil society and democratization. See David Booth (ed.) (1994) *Rethinking Social Development: Theory, Research, and Practice*, Essex: Longman; John Clark (1990) *Democratizing Development: The Role of Voluntary Organizations*, Hartford: Kumarian Press; John Friedmann (1992) *Empowerment: The Politics of Alternative Development*, Cambridge Mass.: Blackwell; Stephen N. Ndegwa (1996) *The Two Faces of Civil Society: NGOs and Politics in Africa*, Hartford: Kumarian Press; M.D. Anisur Rahman (1993) *People's Self-Development*, New Jersey: Zed Books; Matthias Stiefel and Marshall Wolfe (1994) *A Voice for the Excluded: Popular Participation in Development—Utopia or Necessity?*, New Jersey: Zed Books; Marshall Wolfe (1996) *Elusive Development*, New Jersey: Zed Books; Adrian

Leftwich (1996) *Democracy and Development*, Cambridge: Polity Press; Kate Wellard and James G. Copestake (eds) (1993) *Non-Governmental Organizations and the State in Africa: Rethinking Roles in Sustainable Agricultural Development*, New York: Routledge; John Farrington and David J. Lewis with S. Satish and Aurea Miclat-Teves (eds) (1993) *Non-Governmental Organizations and the State in Asia: Rethinking Roles in Sustainable Agricultural Development*, New York: Routledge; Anthony Bebbington and Graham Thiele with Penelope Davies, Martin Prager and Hernando Riveros (eds) (1993) *Non-Governmental Organizations and the State in Latin America: Rethinking Roles in Sustainable Agricultural Development*, New York: Routledge.

CHAPTER 2 BRIDGING LITERATURE GAPS: FRAMING THE PROBLEM OF THE POLITICS OF DEVELOPMENT COOPERATION

1 Richard Hall (1991) *Organizations: Structures, Processes and Outcomes*, 236, New Jersey: Prentice Hall. Hall defines "cooperation" drawing on works from other IOR scholars. For the purposes of this study, I have extended Hall's definition to refer to development interorganizational cooperation.
2 David Leonard, "Analyzing the organizational requirements for serving the rural poor" David Leonard and D.R. Marshall (eds) (1982) *Institutions of Rural Development for the Poor*, 1–38, Berkeley: Institute for International Studies.
3 Charles Mulford (1984) *Interorganizational Relations: Implications for Community Development*, chapter 1, New York: Human Sciences Press; Leon Gordenker and Paul Saunders, "Organization theory and international organization" in Paul Taylor and A.J.R. Groom (eds) (1978) *International Organization: A Conceptual Approach*, 88, New York: Nichols Publishing.
4 Gordenker and Saunders 1978: 89.
5 Mulford 1984: 1–3.
6 L. Julian Efird (1977) "Test of bivariate hypotheses relating characteristics of international organizations to interorganizational relations" *The Emporia State Research Studies*, XXVI, 2 (Fall).
7 Ian Smillie (1995) *The Alms Bazaar*, Ottawa: International Development Research Centre.
8 Michael Bratton (1990) "Non-governmental organizations in Africa: can they influence public policy?" *Development and Change*, 21, 91–2; W.E. Smith, F.J. Lethem and B.A. Thoolen (1980) *The Design of Organizations for Rural Development Projects: A Progress Report*, World Bank Staff Working Paper 375, 15, Washington DC: World Bank.
9 Gordenker and Saunders 1978: 101.
10 Chadwick Alger (1978) "Extending responsible public participation in international affairs" *Exchange*, XIV (Summer), pp. 18–19.
11 Chadwick Alger (1970) "Problems in global organization," paper presented at the Conference on Organization Theory, Comparative Politics and Administration, 22–23 May 1970, City University of New York, as cited in Efird, pp. 6–7.
12 David A. Whetten (1981) "Interorganizational relations: a review of the field" *Journal of Higher Education*, 52, 1, 1.
13 Karen S. Cook (1977) "Exchange and power in networks of interorganizational relations" *Sociological Quarterly*, 18 (Winter), 62–82; Mulford 1984: 4–8.
14 Joseph Galaskiewicz (1979) *Exchange Networks and Community Politics*, Beverly

Hills: Sage; Hall 1991: 10–11. Galaskiewicz and Hall demonstrated the continuity of community-focused IOR studies well into the 1980s.

15 William J. Dixon, "Research on research revisited: another half decade of quantitative and field research on international organizations," unpublished paper cited in Efird 1977: 7.

16 K. Ohene-Bekoe (1990) "Towards globalism in the 21st century: the human factor" *Texas Journal*, 13, 1 (Fall/Winter), 4–7.

17 Chadwick Alger (1979) "The organizational context of development: illuminating paths for wider participation," contribution to the Sub-Project on Non-Territorial Actors of the GPID Project, Japan: The United Nations University, 2–4.

18 *Ibid.*, citing D. H. Smith *et al.*, "Role of US NGOs in international development cooperation" Berhanykun Andemicael (ed.) (1975) *Non-Governmental Organizations in International Co-operation for Development*, New York: UNITAR.

19 Pierre Pradervand (1988) "Self reliance in Africa: peasant groups key to continent's survival" *Christian Science Monitor*, 9 March 1988, section 2, 15–16; Pierre Pradervand (1990) *Listening to Africa: Developing Africa From the Grassroots*, 17–18, New York: Praeger.

20 The term "the North" refers to the industrialized states including most of Western Europe, the United States, Canada, Japan, Australia and New Zealand, most of whom belong to the Organization for Economic Cooperation and Development (OECD) or the "rich man's club." "The North" is sometimes used interchangeably with "the First World" or "the developed countries." The term "the South" refers to the world's less developed or developing states. It includes the rich but yet-to-industrialize oil states of the Middle East and extends to the so-called "Fourth World", the poorest countries on the globe, located for the most part in sub-Saharan Africa, on the Indian sub-continent and in the Caribbean. "The South" is used interchangeably with "less developed countries" (LDCs), underdeveloped countries, developing countries and "the Third World." See Roger D. Hansen (1979) *Beyond the North–South Stalemate*, 3–4, New York: McGraw Hill; *Dictionary of International Relations Terms*, Washington DC: Department of State Library (1987).

21 Gordenker and Saunders 1978: 89; Christer Jonsson (1986) "Interorganization theory and international organization" *International Studies Quarterly*, 30, 39.

22 Jonsson 1986: 41.

23 Several books provide useful discussions of the heavy-handedness of the African state. Among them are Naomi Chazan and Donald Rothchild (eds) (1988) *The Precarious Balance: State and Society in Africa* and Kathleen Staudt and Jane Parpart (eds) (1990) *Women and the State in Africa* (Boulder: Lynne Rienner Publishers).

24 Hall 1991: 216.

25 Amos H. Hawley, "Human ecology" D.L. Sills (ed.) (1968) *International Encyclopedia of the Social Sciences*, 91–3, New York: Macmillan; quoted in Hall 1991: 199.

26 J. C. Mitchell, "The concept and use of social networks" J. C. Mitchell (ed.) (1969) *Social Networks in Urban Situations*, Manchester: Manchester University Press, 2; as quoted in Mulford 1984: 136.

27 Edward O. Laumann, Joseph Galaskiewicz and Peter V. Marsden (1978) "Community structure as interorganizational linkages" *Annual Review of Sociology*, 4, 458; as cited in Mulford 1984: 136.

28 Howard E. Aldrich (1979) *Organizations and Environments*, 281 and 324–5, New Jersey: Prentice Hall.

29 Andrew Van de Ven and Diane L. Ferry (1980) *Measuring and Assessing Organizations*, 299, New York: John Wiley; as cited in Hall 1991: 217.
30 Edward O. Laumann and Franz U. Pappi (1976) *Networks of Collective Action*, 17–20, New York: Academic Press; as cited in Mulford 1984: 136–7.
31 Joseph Galaskiewicz (1985) "Interorganizational relations" *Annual Review of Sociology*, 11, 281.
32 Hall 1991: 222.
33 L. Metcalfe, "Designing precarious partnerships" Paul Nystrom and William Starbuck (eds) (1981) *Handbook of Organizational Design*, 1, 505, New York: Oxford University Press.
34 Julius Nyang'oro (1994) "Reform Politics and the Democratization Process in Africa" *African Studies Review*, 37 (April), 133–49.
35 Governments' and NGOs' interactions in the agricultural sector are examined in Kate Wellard and James G. Copestake (eds) (1993) *Non-Governmental Organizations and the State in Africa: Rethinking Roles in Sustainable Agricultural Development*, New York: Routledge. See also John Farrington and David J. Lewis with S. Satish and Aurea Miclat-Teves (eds) (1993) *Non-Governmental Organizations and the State in Asia: Rethinking Roles in Sustainable Agricultural Development* (New York: Routledge), where the focus rests on country case studies of government–NGO interaction, with particular regard for agriculture projects and/or programs. In Anthony Bebbington and Graham Thiele with Penelope Davies, Martin Prager and Hernando Riveros (eds) (1993) *Non-Governmental Organizations and the State in Latin America: Rethinking Roles in Sustainable Agricultural Development* (New York: Routledge), the authors' foci skew toward issues of cooperation between NGOs, the state and national agricultural research and extension services. They do, however, tangentially draw in caution about NGOs being shrouded in too much donor "love" expressed by the pouring of monies.
36 Anthony Bebbington and John Farrington (1993) "Governments, NGOs and agricultural development: perspectives on changing interorganizational relationships" *Journal of Development Studies*, 29 (January), 199–219; Brian H. Smith (1990) *More Than Altruism: The Politics of Private Foreign Aid*, New Jersey: Princeton University Press; Stephen N, Ndegwa (1996) *The Two Faces of Civil Society: NGOs and Politics in Africa*, Connecticut: Kumarian Press. Another promising study in this area is Julie Fisher (1997) *Nongovernments: NGOs and Governments in the Third World*, Hartford Conn.: Praeger.
37 Ernesto D. Garilao (1987) "Indigenous NGOs as strategic institutions: managing the relationship with government and resource agencies" *World Development*, 15, 113–20.
38 Ryokichi Hirono (1987) "How to promote trilateral cooperation among governments, NGOs and international agencies, in *World NGO Symposium Nagoya Congress October 6–8, 1987, by the Organization for Industrial, Spiritual and Cultural Advancement International (OISCA)*, Japan: OISCA, 77–85.
39 Nigel Twose (1987) "European NGOs: growth or partnership?" *World Development*, 15, 7–10.
40 Smith *et al.* 1980.
41 Martin de Graaf (1987) "Context, constraint or control? Zimbabwean NGOs and their environment" *Development Policy Review*, 5, 277–301.
42 For example, Loysius P. Fernandez (1987) "NGOs in South Asia: people's participation and partnership" *World Development*, 15, 39–50; Kingston Kajese (1987) "An agenda for future tasks of NGOs: views from the South" *World Development*, 15, 79–86.

43 For example, Tim Brodhead (1987) "NGOs: in one year, out the other", in *World Development*, 15, 1–6; Sheldon Annis (1987) "Can small scale development be large scale policy?" *World Development*, 15, 129–34.
44 Ndegwa (1996: 22) agrees and mentions the harassment and suppression of NGOs in Kenya, Uganda, Zimbabwe, Ethiopia, Sudan, Zambia, Tanzania and South Africa.
45 Smillie 1995: 73–8, 194.
46 de Graaf 1987: 277–8.
47 Attempts have been made to classify NGOs. See David Korten (1990) *Getting to the 21st Century: Voluntary Action and the Global Agenda*, Hartford Conn.: Kumarian Press; John Clark (1990) *Democratizing Development: The Role of Voluntary Organizations*, Hartford: Kumarian Press; Matthias Steele and Marshall Wolfe (1994) *A Voice for the Excluded: Popular Participation in Development—Utopia or Necessity?*, New Jersey: Zed Books.
48 Alan Fowler, "NGOs in Africa: naming them by what they are" Kabiru Kinyanjui (ed.) (1985) *Non-Government Organizations (NGOs): Contributions to Development*, proceedings of a seminar held at the Institute for Development Studies, Nairobi, 19 September 1985 (Nairobi: Institute for Development Studies), 7–30.
49 Twose 1987: 9.
50 Goran Hyden (1983) *No Shortcuts to Progress*, 119, Berkeley: University of California Press; Smillie 1995: 60.
51 Helmut K. Anheier, "Indigenous voluntary associations, non-profits, and development in Africa" Walter W. Powell (ed.) (1987) *The Non-Profit Sector: A Research Handbook*, 420–5, New Haven Conn.: Yale University Press; Sylvester J. Ouma (1980) *A History of the Cooperative Movement in Kenya 1908–1978*, 34–68, Nairobi: Bookwise; *A Guide to Women's Organizations and Agencies Serving Women in Kenya*), 286–90, Nairobi: Mazingira Institute (1985; Smillie 1995: 62.
52 "Development NGOs" *Dossier* (Brussels), 104 (July–August 1987), 51.
53 Twose 1987: 9.
54 Many NGOs exist at various levels which include provincial, divisional, locational, as well as the national level. These are important distinctions, though not relevant in this case.
55 Werner Feld, in *International Relations: A Transnational Approach* (1979: 144–5) explains the difference between types of foreign aid. Bilateral aid is aid from one country to another negotiated at the governmental level; multilateral aid is aid from many different countries to one country through a regional or world grouping, such as the World Bank or the United Nations. Private aid is aid provided by private voluntary organizations that are not government-sponsored. For further discussion, see Steven Browne, *Foreign Aid in Practice* (London: Pinter, 1990).
56 Union of International Associations (ed.) (1993–4) *Yearbook of International Organizations*, London: Agreement with the United Nations, Appendix 2.
57 *Ibid.*
58 Mario Padron (1987) "Non-governmental development organizations: from development aid to development cooperation" *World Development*, 15, 69–77.
59 Naomi Chazan *et al.* (eds) *Politics and Society in Contemporary Africa*, 35–70, Boulder: Lynne Rienner Publishers.
60 Peggy Antrobus (1987) "Funding for NGOs: issues and options" *World Development*, 15, 96.

61 Pradervand 1988: 15–6; Bernard J. Lecomte (1986) *Project Aid: Limitations and Alternatives*, Paris: OECD.
62 Helena Gezelius and David Millwood (1991) "NGOs in Development and Participation in Practice" *Transnational Associations*, 5, 278; Hendrik van der Heijden (1990) "Efforts and programmes of NGOs in the least developed countries" *Transnational Associations*, 1, 19.
63 Pradervand 1989; Michael Bratton (1989) "The politics of government—NGO relations in Africa" *World Development*, 17; Karuti Kanyinga and Njuguna Ngethe (1992) *The Politics of Development Space: The State and NGOs in the Delivery of Basic Services in Kenya*, Working Paper 486, Nairobi: Institute for Development Studies.
64 United Nations (1985) *Statement by the Non-Governmental Organizations Attending the Eleventh Ministerial Session of the World Food Council*, Paris, 10–13 June 1985; African NGO's Environment Network (ANEN) (1985) *The African Crisis: Where Do We Go From Here?*, presentation at the United Nations Office of Emergency Operations for Africa/NGO Africa Emergency Consultation Meeting, 22 July, Geneva; Pradervand 1988, 1989.
65 Garilao 1987: 119.
66 Quoted in Pradervand 1988: 15.
67 David J. Steel (1987) "Commemorative lecture" *World NGO Symposium Nagoya Congress October 6–8, 1987, by the Organization for Industrial, Spiritual and Cultural Advancement International (OISCA)*, Japan: OISCA, 139.
68 de Graaf 1987: 277–8.
69 Hendrik van der Heijden 1987: 111.
70 de Graaf 1987: 277, 282–4; Smith *et al.* 1980: 11
71 Helen Allison (1988) "Is foreign aid an asset?" *African Women Quarterly Development Journal*, 1 (March), 18–20.
72 Kabiru Kinyanjui, "The African NGOs in context," presentation at NGOMESA Workshop on the theme "The African NGO Phenomenon: A Reflection for Action" at Gaborone, Botswana, 15–19 May 1989.
73 Sam Kobia, "The old and the new NGOs: approaches to development" Kabiru Kinyanjui (ed.) (1985) *Non-Government Organizations (NGOs): Contributions to Development: Proceedings of a Seminar Held at the Institute for Development Studies, Nairobi, 19 September 1985*, Nairobi: Institute for Development Studies, 35–6.
74 As quoted in Kobia 1985: 35.
75 Yash Tandon, "Foreign NGOs, uses and abuses: an African perspective," excerpt from a paper read at the Round Table organized by the African Association of Public Administration (AAPAM) and the United Nations Economic Commission for Africa/Special Action Program in Administration and Management (ECA/SAPAM) Regional Project on the theme "Mobilizing the informal sector and non-governmental organization for recovery and development: policy and management issues", Abuja, Nigeria, December 1990. The argument presented in this paper is reminiscent of Tandon, "The interpretations of international institutions from a third world perspective" Paul Taylor and A. J. R. Groom (eds) *International Organization: A Conceptual Approach*, 357–80, New York: Nichols.
76 Sithembiso Nyoni (1987) "Indigenous NGOs: liberation, self-reliance, and development" *World Development* , 15 (supplement), 51–6.
77 Korten and Brown 1988: 17–21; Bratton 1989: 14–23.
78 Milton Esman and Norman Uphoff (1984) *Local Organizations: Intermediaries in Rural Development*, 267, Ithaca: Cornell University Press.
79 Bratton 1989: 23–35.

80 Ndegwa 1996: 21–5.
81 *Ibid.*, 6.
82 *Ibid.*, 6–8.
83 *Ibid.*, 25.
84 *Ibid.*, 24.
85 *Ibid.*, 31–54; Smillie 1995: 77.
86 During the writing of this manuscript it was suggested that the works of Tim Brodhead demonstrate the strengths and limitations of NGO activities, yet do not fit either one of these schools. See Brodhead 1987.
87 Pradervand 1988: 15.
88 *Ibid.*; John Clark (1990) *Democratizing Development: The Role of Voluntary Organizations*, Connecticut: Kumarian Press, see pages 64–9 in particular.
89 Shelby Lewis, "African women and national development" Beverly Lindsay (ed.) (1980) *Comparative Perspectives of Third World Women: The Impact of Race, Sex and Class*, 32, New York: Praeger; Barbara Rogers (1980) *The Domestication of Women: Discrimination in Developing Societies*, New York: St. Martin's Press; Ester Boserup (1970) *Woman's Role in Economic Development*, New York: St. Martin's Press.
90 One example is Janet Momsen (1990) *Women and Development in the Third World*, New York: Routledge.
91 Sue Ellen M. Charlton (1984) *Women in Third World Development*, 39, Boulder: Westview Press.
92 At the African Studies Association annual meetings, for instance, the same circle of women present papers and attend sessions, normally about women and gender. In addition, on my return to the field in the summer of 1996, many of this study's interviewees were shocked to see me carrying my research findings back to them for their critique.
93 Charlton 1984: 39.
94 Rogers makes this point somewhat differently. She states, "Male bias has been built into development institutions, processes and policies, and even if all new programs were placed overnight on a foundation of equal access for all, regardless of gender, the momentum of unequal processes already in operation would remain very powerful" (1980: 44).
95 Karl Deutsch (1961) "Social mobilization and political development" *American Political Science Review*, 60, 493–514; Daniel Lerner (1958) *The Passing of Traditional Societies: Modernizing the Middle East*, New York: The Free Press; Gabriel Almond (ed.) (1960) *The Politics of Developing Areas*, New Jersey: Princeton University Press, introduction and chapter 3 by James Coleman.
96 Andre Gunder Frank (1967) *Capitalism and Underdevelopment in Latin America*, New York: Monthly Review Press (Frank draws on the work of Raul Prebisch); Walter Rodney (1971) *How Europe Underdeveloped Africa*, Washington DC: Howard University Press; various works of Samir Amin, particularly "Underdevelopment and dependence in black Africa: origins and contemporary forms" *Journal of Modern African Studies*, 10 (1972), 503–24; Colin Leys (1974) *Underdevelopment in Kenya: The Political Economy of Neocolonialism*, Berkeley: University of California Press; Fernando Cardoso and Enzo Faletto (1979) *Dependency and Development in Latin America*, Los Angeles: University of California Press; Peter Evans (1979) *Dependent Development*, New Jersey: Princeton University Press.
97 The specific Wallerstein work referred to here is Joan Smith, Immanuel Wallerstein and Hans-Dieter Evers, (eds) (1984) *Households and the World Economy*, Beverly Hills: Sage. It is important to note that Wallerstein is not a

dependency theorist. He has, however, been influenced by underdevelopment and dependency theory. He has also contributed to the theory's growth. The book cited here has been criticized for dealing with gender very tangentially.

98 Chazan *et al.* (1988) *Politics and Society in Contemporary Africa* (Boulder: Lynne Rienner Publishers), especially chapters 2 and 9; most of the works in Naomi Chazan and Donald Rothchild (eds) (1988) *The Precarious Balance: State and Society in Africa*, Boulder: Westview Press; Parpart and Staudt 1990: 3–7.

99 Samuel Huntington (1968) *Political Order in Changing Societies*, New Haven: Yale University Press; Gunnar Myrdal in *Asian Drama: An Inquiry into the Poverty of Nations* (New York: Pantheon, 1968) coined the concept "soft state;" Leonard Binder *et al.* (1971) *Crises and Sequences in Political Development*, New Jersey: Princeton University Press, as cited in Milton Esman and Norman Uphoff (1984) *Local Organizations: Intermediaries in Rural Development*, Ithaca: Cornell University Press.

100 Many of the political economy strategies are discussed in Frederick Deyo (ed.) (1987) *The Political Economy of the New Asian Industrialism*, Ithaca: Cornell University Press; Wayne McWilliams and Harry Piotrowski (1993) *The World Since 1945*, chapter 13, Boulder: Lynne Rienner Publishers.

101 Debbie Taylor *et al.* (1985) *Women: A World Report*, 29–41, New York: Oxford University Press.

102 R. William Liddle (1992) "The politics of development policy" *World Development*, 20 (June), 793–807.

103 Peter Ekeh (1975) "Colonialism and the two publics in Africa: a theoretical statement" *Comparative Studies in Society and History*, 17, 91–112. Ekeh identifies the primordial public as that public identified with primordial groupings and sentiments. It is moral and operates on the moral imperatives of the private realm. The civic public is identified with civil structures associated with colonial administration. It is amoral and lacks the moral imperative of the private realm.

104 Pearl Robinson proposes that there exists a culture of politics that is "culturally legitimated and societally validated by local knowledge." I agree and would argue that it is also gendered. See Robinson (1994) "Democratization: understanding the relationship between regime chance and culture of politics" *African Studies Review*, 37, (April), 39–40.

105 Kathleen Staudt and Harvey Glickman (1989) "Beyond Nairobi: women's politics and policies in Africa revisited" *Issue: A Journal of Opinion*, 17 (Summer), 4–6. 1992 interview. She discussed male bias as a primary issue.

106 Catherine V. Scott (1995) *Gender and Development: Rethinking Modernization and Dependency Theory*, 4, Boulder: Lynne Rienner Publishers.

107 *Ibid.*, 5–6.

108 *Ibid.*, 6.

109 *Ibid.*, 19.

110 Jane L. Parpart, "Post-modernism, gender, and development" Jonathan Crush (ed.) (1995) *Power of Development*, 258–62, New York: Routledge. WAD represents an approach to development inspired by dependency theorists and radical feminist theorists such as Carol Gilligan in the 1980s. Parpart states: "They called for self-reliant development, free from the self-interested 'assistance' of capitalist elites and their indigenous henchmen." GAD represents a more recent shift in analytical thinking from analyses based on women's conditions to gender relations. GAD does not focus on women in isolation, but instead on their relations to men.

111 *Ibid.*, 259.

112 *Ibid.*, 264.
113 *Ibid.*, 261.
114 David Booth (1994) *Rethinking Social Development: Theory, Research, and Practice*, Essex: Longman Scientific and Technical.
115 Caroline Moser (1993) *Gender Planning and Development: Theory, Practice, and Training*, introduction, New York: Routledge.
116 *Ibid.*, 196–8. Moser's work builds on that of Development Alternatives with Women for a New Era (DAWN) from an earlier decade, yet DAWN's work departs fundamentally from that of Moser, in that less of their focus is on gender struggle and equity, equality and empowerment issues. DAWN's position is that development solutions must be addressed through global redistribution.
117 John Saul, in his article "The role of ideology in transition to socialism" R. Fagen (ed.) (1986) *Transition and Development: Problems of Third World Socialism*, New York: Monthly Review Press. Parpart and Staudt (1990: 1) employ this argument and cite Saul. However, they replace "man" with "woman."
118 Robert Fatton, "Gender, class, and state in Africa" Parpart and Staudt (1990: 48–9).
119 Akosua Adomako Ampoto (1993) "Controlling and punishing women: violence against Ghanaian women" *Review of African Political Economy*, 56, 103–11. Claire Robertson also makes this argument in *Sharing the Same Bowl: A Socioeconomic History of Women and Class in Accra, Ghana* (Bloomington: Indiana University Press, 1984: 244). Misogynist statements cited were spoken during the destruction of the Makola No. 1 Women's Market in Accra on 18 August 1979. This market had been established in 1924 as the center of trade in Ghana. In 1978, before the destruction, it was chief among 19 markets in Accra, controlled by women traders.
120 Maria Nzomo (1991) "Women as men's voting tools" *Kenya Times*, 4 September 1991, 14–15.
121 Charlton 1984: 24.
122 John Stoltenberg, "Toward gender justice" Jon Snodgrass (ed.) (1977) *For Men Against Sexism*, 75–6, Californiia: Times Change Press; quoted in bell hooks (1981) *Ain't I A Woman? Black Women and Feminism*, 99–100, Boston: South End Press.
123 bell hooks, "Men: comrades in struggle" bell hooks (ed.) (1984) *Feminist Theory from Margin to Center*, 67–81, Boston: South End Press.
124 Ibid.; see also all hooks 1981, 1984.
125 Shelby Lewis, "A liberationist ideology: the intersection of race, sex, and class" Molly Shanley (ed.) (1988) *Women's Rights, Feminism, and Politics in the United States*, 38–44, Washington DC: American Political Science Association.
126 Nzomo 1991a: 115.
127 Staudt and Glickman 1989: 5, drawing on research by Bonnie Keller in Zambia.
128 Nzomo 1989: 15–6.
129 Mike Savage and Anne Witz (eds) (1992) *Gender and Bureaucracy*, 3, Cambridge: Blackwell.
130 Celia Davies, "Gender, history and management style in nursing: towards a theoretical synthesis" Savage and Witz 1992: 229–52; Savage and Witz 1992: 26.
131 Karen Ramsay and Martin Parker, "Gender, bureaucracy and organizational culture" Savage and Witz 1992: 259–60.

132 Rosemary Pringle (1989) *Secretaries Talk: Sexuality, Power and Work*, 47 and 92, New York: Verso, as cited in Savage and Witz 1992: 27.
133 Ramsay and Parker in Savage and Witz 1992: 259.
134 Elise Boulding (1975) "Female alternatives to hierarchical systems, past and present" *International Associations*, 6–7, 340–6.
135 Pradervand 1988, 1989; Parpart and Staudt 1990: 8.
136 Parpart and Staudt 1990: 8; Chazan in Parpart and Staudt 1989: 185–201.
137 Nzomo 1989: 15; Marjorie Mbilinyi (1984) "Women in development ideology: the promotion of competition and exploitation" *The African Review*, 2, 14–33.
138 Rogers 1980, entire work.
139 Sally Yudelman (1987) "The integration of women into development projects: observations on the NGO experience in general and in Latin America in particular" in *World Development*, 15, 186.
140 Staudt and Glickman 1989: 5.
141 Fatton in Parpart and Staudt 1990: 54.
142 Nzomo 1991b: 28–9.
143 Kathleen Staudt (1980) "Women's organizations in rural development" (report distributed by the Office of Women in Development, US Agency for International Development, Washington DC), 29, as quoted in Charlton 1984: 211.
144 Janet Bujra, "Urging women to redouble their efforts: class, gender and capitalist transformation in Africa" in Fatton 1986: 63.
145 For a brief discussion see Parpart and Staudt 1990: 9–11.
146 Mulford 1984: 83.
147 Hall 1991: 277–82.
148 Sol Levine and Paul E. White (1960–1) "Exchange as a conceptual framework for the study of interorganizational relations" *Administrative Science Quarterly*, 5, 583.
149 Mulford 1984: 79.
150 Karen Cook (1977) "Exchange and power in networks of interorganizational relations" *Sociological Quarterly*, 18, 1 (Winter), 63–4.
151 Ibid.; Mulford 1984: 81–3; Aldrich 1979: 267.
152 Aldrich 1979: 267.
153 Mulford 1984: 47, 83.
154 Richard Emerson (1962) "Power–dependence relations" *American Sociological Review*, 27 (February), 32.
155 Ronald Burt (1977) "Power in a social typology" *Social Science Research*, 6, 1–83, cited in Mulford 1984: 47.
156 This scenario was gleaned from readings which cover the scope of IOR theory. Particularly suggestive was Charles Mulford's *Interorganizational Relations* (1984) previously cited.
157 Pradervand 1988, 1989.
158 Robert Dahl (1957) "The concept of power" *Behavioral Science*, 2 (July), 201–15; Emerson 1962: 31–40.
159 Peter Blau (1964) *Exchange and Power in Social Life*, 118, New York: John Wiley, as cited in Mulford 1984: 84.
160 William Evan (1976) *Organization Theory: Structures, Systems and Environments*, 148–66, New York: John Wiley and Sons; Christer Jonsson (1986) "Interorganization theory and international organization" *International Studies Quarterly*, 30, 39–57.
161 Evan 1976: 149–150.

162 Talcott Parsons (1960) *Structure and Process in Modern Societies*, Illinois: The Free Press.
163 Ibid.; Evan 1976: 149.
164 Jonsson 1986: 42.
165 *Ibid.*, 43.
166 Howard Aldrich and David Whetten, "Organization sets, action sets, and networks: making the most of simplicity" in Paul Nystrom and William Starbuck (eds) *Handbook of Organizational Design*, 390, New York: Oxford University Press.

CHAPTER 3 THE EVOLUTION OF MYWO FROM 1952 TO 1992

1 Audrey Wipper (1975–6) "The Maendeleo Ya Wanawake movement in the colonial period: the Canadian connection, Mau Mau, embroidery and agriculture" *Rural Africana*, 29 (Winter), 197–201.
2 *Ibid.*; Audrey Wipper (1975) "The Maendeleo Ya Wanawake Organization: the co-optation of leadership" *African Studies Review*, 18 (December), 100.
3 Cora Presley (1992) *Kikuyu Women, the Mau Mau Rebellion, and Social Change in Kenya*, 166, Boulder: Westview Press.
4 *Ibid.*
5 Wipper 1975.
6 Wipper 1975: 99–101; Charles Otieno (1982a) "A history of Maendeleo" *Viva* (Nairobi), 9, 11–17, 88–90.
7 Otieno 1982a: 13.
8 *Ibid.*
9 *Ibid.*
10 Public Record Office (1952b) "Maendeleo Ya Wanawake: Extract from the Federation of Social Services Report for 1952" ; Wipper (1975) indicates that the Department of Community Development (DCD) suggested the following program for weekly meetings over a three-month period:

 1 bathing a baby
 2 health and hygiene in the home
 3 agriculture—rotation of crops
 4 children's play, training in character building
 5 clothing—choice of suitable clothes for climate, etc.
 6 how to build a mud stove (if the women have their own house)
 7 child welfare
 8 hygiene and health in the home
 9 recipes or cooking demonstrations
 10 agriculture—compost and compost pits
 11 handiwork or needlework demonstration
 12 tea party and concert
 13 talk on current affairs
 14 literacy classes should be held in addition to the club meeting.

11 Public Record Office, Colonial Government of Kenya (1952a) *Community Development Organization—Kenya: Establishment and Annual Report for 1952*; Otieno 1982a: 13–15.
12 *PRO 1952a.*
13 Otieno 1982a: 13.
14 *Ibid.*, 15.

15 *Ibid.*, 13.
16 PRO (1957) *History of Mau Mau Emergency in Kenya*. The propaganda campaign is also referred to as "Hammer" in some of the records.
17 Carl Roseberg and John Nottingham (1966) *The Myth of "Mau Mau": Nationalism in Kenya*, New York: Praeger; Robert Buijtenhuis (1973) *Mau Mau Twenty Years After: The Myth and Survivors*, Mouton: The Hague Press; B.A. Ogot (1972) *Politics and Nationalism in Colonial Kenya*, Nairobi: East African Publishing House; Sam Kahiga (1990) *Dedan Kimathi: The Real Story*, Nairobi: Longman.
18 Presley 1992; Tabitha Kanogo (1987) *Squatters and the Roots of Mau Mau 1905–1963*, Columbus: Ohio University Press; Colin Legum (1953) "Mau Mau Organization: psychological warfare needed" *The Scotsman*, 12 August 1953; PRO (1953b) *Psychological Warfare in Kenya*.
19 *Ibid.*
20 PRO (1953d) "Notes from the Working Party on Information"; PRO (1953b) *Psychological Warfare in Kenya*.
21 Presley 1992: 166.
22 PRO (1954–5) "Women's Clubs are 'Valuable Rallying Points'" extract from Kenya Newsletter 85.
23 *Ibid.*
24 Otieno 1982a: 15; Maria Nzomo (1992b) "Schemes to divide and oppress women started in the colonial era" *The Sunday Nation*, 17 May 1992, 17 (Nairobi).
25 PRO (1955c) "Women's Club Drive in Kiambu" Press Office Handout 621, 7 June 1955.
26 *Ibid.*; PRO (1955b) "The Lady Mary Baring Visits Kikuyu Women's Clubs" Press Office Handout 207, 24 February 1955.
27 *Ibid.*
28 *Ibid.*; PRO (1953e)"Demand for Women's Clubs in Kiambu" Press Office Handout 840, 17 December 1953.
29 *Ibid.*; PRO (1954d) "Lady Baring Machakos Safari" Press Office Handout 1351, 28 October 1954.
30 *Ibid.*; PRO (n.d.) "Women's Clubs Great Progress in Machakos" Press Office Handout 47.
31 PRO (1952c) "Work in Districts" *Community Development Organization Annual Report for 1952*.
32 PRO (1954c) "European Women Build Club for African Women" Press Office Handout 1331, 22 October 1954.
33 *Ibid.*; PRO (1955d) "Classes for Kitale African Women" Press Office Handout 965, 23 August 1955.
34 *Ibid.*; PRO (1954b) "African Women Complete Domestic Science Course" Press Office Handout 1044, 12 August 1954.
35 *Ibid.*; PRO (1955e) "Over 40,000 members of African Women's Clubs" Press Office Handout, 24 August 1955.
36 *Ibid.*; PRO (n.d.) "African Women's Clubs" extract from Kenya Newsletter 90.
37 PRO (1955a) "Progress Report".
38 PRO (1958) "Letter from Mrs Prefumo to Miss Joan Vickers" 5 August 1958.
39 Wipper 1975–6: 197.
40 Wipper 1975: 102.
41 PRO (1953b) "Psychological Warfare" file. The "classes" of persons to whom this propaganda campaign was targeted among Africans were:

1 the Loyalists to the colonial government;
2 the Waverers, who might be responsive to Mau Mau but not totally committed;
3 the Mau Mau "Gangsters";
4 those involved in development and reconstruction projects, including MYWO club women.

42 *Ibid.*
43 PRO (1954a) "Maendeleo Ya Wanawake: Extract from the Federation of Social Services Annual Report 1954".
44 PRO (1956) "Terrorist Strength" January 1956 estimates, in *Future Emergency Policy in Kenya*.
45 PRO (1954a) "Maendeleo Ya Wanawake Extract from the Federation of Social Services Annual Report 1954".
46 PRO (1953c) "Psychological Warfare to All".
47 *Ibid.*; and PRO (1953b) "Psychological Warfare", section headed "Information on the Kikuyu, Meru and Embu in the Rift Valley Province and the Settled Areas of Central Province".
48 PRO (1953b) "Psychological Warfare."
49 PRO (1955a) "Progress Report."
50 Otieno 19982a: 17.
51 *Ibid.*, 15–17.
52 *Ibid.*, 88.
53 Otieno (1982b) "Past chairmen" *Viva* (Nairobi), 9, 21.
54 Otieno 1982a: 88.
55 *Ibid.* One of MYWO's programs was a literacy program in which one literate member of the organization would teach at least one illiterate member. Another of their programs focused on homecraft, particularly in rural areas, making it income-generating. Their third program was the revival of the oral tradition—story-telling and oral literature.
56 Otieno 1982a: 89.
57 *Ibid.*
58 *Ibid.*
59 *Ibid.*
60 Elspeth Huxley (1968) *White Man's Country: Lord Delamere and the Making of Kenya*, New York: Praeger.
61 Otieno (1982c) "An Interview with Mrs Kiano" *Viva* (Nairobi), 9, 28–34, 85–7.
62 Nzomo 1992b: 17.
63 *Ibid.*
64 See David Himbara (1994) *Kenyan Capitalists, the State and Development*, 27–8, 94–5, Boulder: Lynne Rienner Publishers; and "Renewed calls of Majimboism" *Weekly Review* (Nairobi), 13 September 1991, 5–12, for a discussion of members and impact of GEMA on Kenya's politics.
65 For the shift to rural focus, see Joel Barkan and Michael Chege (1989) "Decentralizing the state: district focus and the politics of reallocation in Kenya" *Journal of Modern African Studies*, 27, 3. For the bias against women, see Maria Nzomo (1991a) "Women as men's voting tools" *Kenya Times*, 4 September 1991; and Maria Nzomo (1991b) "Women's passivity to blame for their woes" *Kenya Times*, 8 September 1991.
66 Otieno 1982b: 23.
67 Otieno 1982a: 89.

68 Otieno 1982b: 23.
69 Otieno 1982a: 90.
70 Otieno 1982b: 23.
71 Otieno 1982a: 90.
72 1992 interview.
73 Otieno 1982e: 45; *A Guide to Women's Organization and Agencies Serving Women in Kenya*, 288, Nairobi: Mazingira Institute, 1985.
74 Maria Nzomo (1989) "The impact of the Women's Decade on policies, program and empowerment of women in Kenya" *Issue: A Journal of Opinion*, 7, 9–17; "Wangari Maathai" *Presence*, 8 (January 1992), 3–6, 12–14, 20, 30. Neither article mentions Kiano by name, but by reference to her chairship the authors assess her impact on women in Kenya. Wipper (1975) mentions Kiano by name.
75 Nzomo 1992b: 17; Otieno 1982e: 41; Otieno Onyango (1992) "The Terminal Crisis for Maendeleo" *Presence*, 7, 18.
76 Otieno 1982b: 23.
77 Otieno (1982f) "Committee members" *Viva* (Nairobi), 9, 63.
78 Jane Kiano (1982) "What we stand for" *Viva* (Nairobi), 9, 7 and 73.
79 Otieno 1982f: 63. Another interviewee in 1992 indicated that during Kiano's chairship it was a national obligation to become involved with development and with women. "Work was a must. . . . The nation had to be built." She and Otieno recounted how Kiano had grown up in a tradition of assisting women—her sister, Gladys, aiding widowed women in their home areas, and she and her school friends helping older and more needy women. Thus, Kiano continued the tradition of working with women and called on all women to respond in the spirit of *harambee*.
80 1992 interview.
81 *Ibid.*
82 For a discussion of Kenyatta's "ethnic/tribal politics", see Himbara 1994: 94–5 and 118–19.
83 *Ibid.*
84 1992 interview.
85 *Ibid.*
86 To fully understand the importance of unofficial exchanges one must know how tenuous the position of Indian entrepreneurs was in the 1970s. As Himbara argues in *Kenyan Capitalists* (1994: 59–65), Indians were often intentionally displaced by the policies of Africanization formulated and implemented in the early post-independence period. The House of Manji was not financially affected, though many other businesses were. Over time, however, after most African would-be entrepreneurs had failed, many Indians were able to return, with an even stronger economic base.
87 1992 interview.
88 Conversation with Dr Gikonyo Kiano, 1991; Emily Onyango (1984) "Matiba, Kiano Make Up at Last" *Kenya Times*, 6 June 1984.
89 1992 interview.
90 *Ibid.*
91 Otieno (1982d) "Maendeleo and the women of the world" *Viva* (Nairobi), 9, 37.
92 1992 interview.
93 *Ibid.*
94 Wipper 1975: 114–16.
95 John Orora and Hans Spiegel (1980) "Harambee: self-help development

project in Kenya" *International Journal of Comparative Sociology*, XXI, 3–4, 243–53.

96 *Ibid.*

97 Grace N. Wanyeki (1985) "Report from Central Province" *Women's Voice: Official Journal of Maendeleo Ya Wanawake*, 3, 16.

98 Otieno 1982c: 33–4.

99 1991 interview.

100 1989 interview; 1992 interview.

101 1992 interview.

102 Otieno 1982c: 29.

103 1992 interview.

104 1992 interviews with former MYWO officials from the 1970s and 1980s, revealed that their perception was that MYWO members created their own development agenda.

105 Mayengo Mukhwana (1985) "Cooperate with the government" *Women's Voice*, 3 (April), 1.

106 Kenya Ministry for Culture and Social Services, Women's Bureau (1985) *Women of Kenya: Review and Evaluation of Progress*, 49, Nairobi: Kenya Literature Bureau.

107 Nzomo 1992b: 17; Nzomo 1989: 9.

108 Nzomo 1989: 9.

109 Damaris Kamau (1985) "Report from the Thika Township Women Group" *Women's Voice*, 3 (June), 17.

110 Republic of Kenya (1973) "Sessional Paper No. 10 on Employment", 64, Nairobi: Government Printer.

111 For an analysis of the position of the government see Nzomo 1989: 9.

112 Orora and Speigel 1980: 244–6.

113 *Ibid.*, 249–50.

114 *Ibid.*, 243.

115 Republic of Kenya (1989) *Development Plan 1989–1993*, 258–9, Nairobi: Government Printer.

116 *Ibid.*, 259–60.

117 Victoria Okumu (1983) "Oloitipitip advises women to raise cash" *Kenya Times*, 16 April 1983.

118 Daniel T. arap Moi (1986) *Kenya African Nationalism: Nyayo Philosophy and Principles*, 118, London: Macmillan.

119 *Ibid*, 116.

120 *Ibid*, 118.

121 *Ibid.*

122 Nzomo 1989: 15; Nzomo (1991) "Women's passivity to blame for their woes" *Sunday Times,* 8 September 1991.

123 1991 interview.

124 Onyango 1992: 19.

125 Otieno 1982a: 90.

126 Wahome Muthai (1991) "Attitudes will change if women elites stand firm" *Daily Nation*, 23 September 1991.

127 Nzomo 1989: 15; Marjorie Mbilinyi (1984) "Women in development ideology: the promotion of competition and exploitation" *The African Review*, ll, 1, 14–33. Nzomo and Mbilinyi argue that foreign donors are the ones who exploit women's labor.

128 "Sh 40 million for KMYWO projects" *Nation*, 2 July 1990.

129 Hussein Mohammed (1991) "P.S. says Kenya has Sh 154 billion debt" *Kenya Times*, 19 November 1991.
130 1991 interview.
131 Onyango 1992: 19; "Maendeleo in financial mess" *Standard*, 10 November 1985.
132 "Maendeleo to be probed" *Standard*, 11 November 1985.
133 *Ibid*; Onyango 1992: 19.
134 "Now Maendeleo top officer is suspended" *Standard*, 12 December 1985.
135 "Curb on use of vehicles" *Standard*, 21 December 1985.
136 The MCSS probe report was released on 14 February 1986. MYWO was threatened with lawsuits from merchants who had tried unsuccessfully to collect payments for debts which MYWO had incurred. These included:

Name	*Ksh*
Dharamshi Devraj Shah	247,300.00
Munshiran International Business Machines	40,103.25
Pan Plastics Ltd	33,330.00
Menegai Furniture	21,235.00
Gut Ltd	7,408.80

The MCSS probe found that:

1 MYWO's financial undertakings and procurement procedures were in need of an overhaul. Most of MYWO's financial transactions had not followed the constitutional stipulations of necessary sanctioning by the financial or national executive committee;
2 Shitakha had, in fact, practiced nepotism in contracting her relative Mr Matthew Shitakha of East West Communications for the publication of the April, May and June editions of the *Women's Voice*, the official journal of MYWO. There had been no other bids solicited from other publishers. This deal between Shitakha and Shitakha cost MYWO Ksh 165,000, whereas formerly the journal is reported to have been self-supporting from advertisements. East West Communications had pocketed Ksh 142,000 collected in advertising revenue. East West Communication were further awarded another 115,000 by MYWO. Above all, they were not even publishers;
3 MYWO had used improper procurement worth Ksh 1.449 million for the Decade of Women Conference;
4 MYWO had unaccounted sales of Ksh 78,858.15 and an outstanding imprest of Ksh 61,000;
5 MYWO had improperly managed vehicles;
6 MYWO had run bank accounts poorly;
7 MYWO irregularly made appointments on tribal grounds.

"Women body faces legal suits" *Standard*, 6 February 1986; Onyango 1992: 19.
137 Catherine Gicheru (1986) "Maendeleo boss gets the sack" *Daily Nation*, 14 February 1986; Nelson Osiemo (1986) "Maendeleo boss sacked" *Standard*, 14 February 1986.
138 Onyango 1992: 19.
139 Odongo Odoyo (1985) "MYWO looks to district focus" *Standard*, 12 November 1985.
140 1992 interview.
141 Onyango 1992: 19.

142 KANU/MYWO Working Committee, Minutes of 3 March 1987, Third Meeting on KANU/MYWO Merger at KANU Headquarters, reference number MYWO/KANU/12/3; KANU/Maendeleo Ya Wanawake Organization Constitution and Rules, Objective Section, 4.

143 The caretaker committee and the KANU/(K)MYWO working group found that MYWO owed Ksh 5, 974,655 in taxes to the Commissioner of Income Tax. In attempts to resolve this enormous tax problem, the caretaker committee wrote to MCSS in an appeal to the Minister of Finance requesting that MYWO be exempted from paying these taxes. The Minister of Finance agreed to waive the accumulated taxes, but ordered that the late payment penalties of Ksh 821,003 be made. This waiver did not include income taxes of 1985–6 totaling Ksh 3.5 million, as they were not assessed at the time of the negotiations. MYWO also had a host of creditors which it had not paid for "bills incurred in 1985 during the UN End of the Women's Decade Conference." These creditors included, but are not exclusive to, the ones aforementioned. They follow:

Name	*Ksh*
Nation newspaper	2,358.00
Furniture Centre	7,150.00
Biuk Electrical Ltd	1,000.00
Menegal Furniture	21,235.00
Pan Plastics Ltd	33,330.00
Maanki Auto Garage	23,785.00
Rex Rotary	5,340.00
Rentokil	560.00
Dharamshi Devraji Shah	50,000.00
Gut Ltd	7,408.00
Kenya Times	4,968.00
City Council of Nairobi	7,556.05
GTZ	300,000.00
Republic of Kenya	7,275.00
Munishram International	40,103.25
Genuine Merchants	38,000.00
Grace Ngethe	14,700.00
Computer Typesetters	6,200.00
Copy Cat	5,600.00
MCH/FP	94,465.00
KP&T	6,406.40
Keni Office	2,229.95
Tarpo Industries	15,750.00
East West Communication	100,000.00
Dharamshi Devraji Shah	197,300.00
Design Directions	79,650.00
Time Joints	300,000.00
African Insurances	18,696.00
Water A/C	33,668.00
Minanzo Ltd	3,053.00
Vincent Mugemi	4,200.00
Total Ksh	1,431,989.00

Overall liabilities totaled Ksh 11,762,990.00

144 According to the constitutional changes:

1 **MYWO's title was changed to KANU MYWO (KMYWO).** MYWO ceased being a non-political NGO and became part of a political party and the state.
2 **KMYWO was declared the official voice and coordinator of all women's organizations in the country.** This placed KANU above and in charge of all women's organizations, and all of their finances.
3 **KMYWO would be disciplined by the National Disciplinary Committee of KANU, subject to its rules and regulations.** Hence, any member of KMYWO perceived as bringing "disrepute" to the organization or party would be disciplined. That is, women could not speak out against KANU and would be further silenced against challenging the Moi government and the state.
4 **KMYWO committees would have additional voting members, each committee would have 1 voting KANU representative, with the exception of the Annual General Meeting.** KMYWO's policies, practices, agenda and "women's voice" could be affected.
5 **KMYWO would prepare and submit to the KANU National Executive Committee a financial statement every quarter.** KANU would be consistently privy to MYWO purse strings.
6 **KMYWO would submit to the KANU National Executive Committee quarterly reports on all overseas funding.** KANU would be privy to all financial negotiations and donations from foreign donors to KMYWO. KANU would moreover have the power to impact the relationship between KMYWO and foreign donors, as KANU now had the right to vote in committees that made decisions about donors, funding and projects.
7 **KMYWO would pay affiliation fees only to other national or international associations approved by KANU.** KANU could now determine the involvement of KMYWO in women's and other activities worldwide. It could restrict KMYWO's growth and involvement.
8 **KMYWO would have a KANU representative present at KMYWO's Annual General Meetings to observe.** KANU could now "watch" first hand what the women were up to and censor their planning for the next year.
9 **KMYWO would now elect at every General Meeting a KANU representative to be a trustee of KMYWO.** KANU would have a vested interest in all assets acquired by MYWO, and would be a beneficiary if those assets became profit-making.
10 **KMYWO's constitution would be amended only in consultation with KANU.** MYWO would not change in structure or function without the said permission of KANU, of men.
11 **Women running for office in KMYWO would be registered members of both MYWO and KANU.** KANU would gain the memberships and membership dues of women who, apart from being members of MYWO, might not join KANU.
12 **KMYWO would change its traditional voting system from secret ballot to queuing.** KANU could now see who individuals and groups of women were voting for and hold them accountable. They could also influence who the women queued behind, and thus voted for. As Nzomo indicates, "a husband could then successfully order his wife and other members of his family not to line up behind a female candidate."

145 Jeanette Fregulia and Victoria Okumu (1984) untitled article about Jane Kiano in *Standard*, 29 April 1984.
146 1989 interview.
147 1992 interview.
148 KANU/MYWO, Budget and Program for Proposed KANU Maendeleo Ya Wanawake Workshop, 18 March 1987.
149 Wanjiku Mbugua (1989) "Women's body merges with ruling party" *Standard*, 10 October 1989.
150 *Ibid.*
151 *Ibid.*
152 "KANU and Maendeleo Ya Wanawake—a marriage full of promise" *Viva* (Nairobi), V, 7, 11.
153 *Ibid.*
154 *Kenya Times*, 24 August 1989; *Daily Nation*, 22 September 1989; *Standard*, 22 September 1989, 14 October 1989, 28 October 1989.
155 George Munji (1989) "Maendeleo elections postponed" *Kenya Times*, 24 August 1989; "Maendeleo Ya Wanawake male dominated squabbling" *Weekly Review* (Nairobi), 27 October 1989, 14–15.
156 Emman Omari (1989) "Maendeleo polls postponed again" *Daily Nation*, 22 September 1989.
157 "Male-dominated squabbling" *Weekly Review*, 27 October 1989.
158 "An impressive turnout" *Weekly Review*, 3 November 1989, 8.
159 *Kenya Times*, 24 August 1989; "Male-dominated squabbling" *Weekly Review*, 27 October 1989.
160 "Maendeleo polls postponed again" *Daily Nation*, 22 September 1989; "MYWO elections postponed once again" *Standard*, 22 September 1989.
161 "To the women we say: vote wisely" *Daily Nation*, 30 October 1989.
162 "Mutisya tips on elections" *Daily Nation*, 18 September 1989.
163 *Ibid.*
164 "Maendeleo polls" *Daily Nation*, 19 September 1989.
165 Joel Kipsongok (1989) "Don't be cheated, Moi tells women" *Standard*, 31 October 1989.
166 "President's new rule on Maendeleo elections" *Daily Nation*, 27 October 1989.
167 "Maendeleo polls—a male affair?" *Weekly Review*, (Nairobi), 3 November 1989, 4
168 Haroun Wandal and bureaux reporters (1989) "Men meddle in Maendeleo polls" *Standard*, 31 October 1989.
169 "KANU bosses' wives win Maendeleo polls" *Standard*, 1 November 1989.
170 *Ibid.*
171 "Maendeleo polls—a male affair?" *Weekly Review*, 3 November 1989.
172 "KANU bosses' wives win Maendeleo polls" *Standard*, 1 November 1989.
173 "Men meddle in Maendeleo polls" *Standard*, 31 October 1989.
174 "KANU—Maendeleo elections surprise outcome" *Weekly Review* (Nairobi), 10 November 1989, 10.
175 *Ibid.* From Central Province, Muranga's and Kirinyaga's district chairmen, Joseph Kamotho and James Njiru, with Kipipiri MP, Nyandarua District, Mr Kabingu Muregi, persuaded the Waruhiu to accept the delegation as already recommended—the vice chairship going to Nyeri District and two provincial seats going to Nyandarua and Kirinyaga Districts. She accepted reluctantly. It was rumored that Waruhiu herself may have had her eye on one of the two top offices. Perhaps her difficulty rested in that she was not a hand-picked KANU candidate, although she was a member of the party. And perhaps, as Kiambu

had been the political center of the Kenyatta power base and of GEMA, Kikuyus from Kiambu were not accustomed to being anything but central to the political organs of the day. Moi had tried to replace and annihilate them. Waruhiu may have seen this as another tribal blow against Kiambu political power, and a trick played on her by KANU. See Himbara (1994) *Kenyan Capitalists* for a discussion of the consolidation and collapse of Kikuyu power and the rise of the Kalenjin.

176 "KANU—Maendeleo elections surprise outcome" *Weekly Review*, 10 November 1989.

177 "Maendeleo polls–a male affair?" *Weekly Review*, 3 November 1989.

178 "KANU—Maendeleo polls surprise outcome" *Weekly Review*, 10 November 1989.

179 Mumbi Risah (1989) "Women's noble efforts to develop the nation" *Standard*, 10 October 1989.

180 1993 interview.

181 Elections at the sub-national grassroots levels looked different from elections at the national level. KANU presence and influence was much more obvious. KANU wives and relatives took 20 seats in KMYWO. They included, but were not exclusive to:

Mrs Eunice Kamotho who took the KMYWO Kangema chair, Muranga District. She is the wife of Joseph Kamotho, the then KANU Secretary General and Minister for Transportation and Communications;

Mrs Mary Kamuyu who took the KMYWO Dagoretti Division chair, Nairobi. She is the wife of Chris Kamuyu, MP and KANU sub-branch chair.

Mrs Isabel (Ezabel) Mwenja who took the KMYWO Embakasi Division chair, Nairobi. She is the wife of David Mwenje, MP Embakasi.

Mrs Clare Omanga who took the KMYWO Kisii District chair and a Kisii delegate seat for Borabu Division. She is the wife of Andrew Omanga, MP for Nyaribari Chache and former Minister for Tourism.

Mrs Mary Sagini who took a Kisii delegate seat for Borabu Division. She is the wife of Lawrence Sagini, Kisii KANU branch chair.

Mrs Rebecca Wanjiru Ndung'u who took the KMYWO Kigio Sublocation chair. She is the wife of George Mwicigi, former Assistant Minister for Energy and Regional Development.

Mrs Margaret Ndang'a Mundia who took the KMYWO Thika Division assistant secretary seat. She is the wife of Douglas Mundia, Mayor Councilor of Thika Division.

Mrs Aisha Sharrif Taib Busaidy who took the KMYWO Majengo Division chair, Mombasa District. She is the sister of Shariff Nassir, Assistant Minister for Information and Broadcasting.

See the following: "Men meddle" *Standard,* 31 October 1989; "KANU bosses' wives win Maendeleo polls" *Standard*, 1 November 1989; "Excitement, protests mark Maendeleo polls" *Daily Nation*, 31 October 1989; "Politicians' wives in polls triumph" *Daily Nation*, 1 November 1989; "Maendeleo fresh polls call" *Standard*, 17 June 1990.

182 *Ibid*; Patrick Wakhisi (1991) "Maendeleo branch Members want polls cancelled" *Standard*, 1 February 1991.

183 "Taking the bull by the horns" *Weekly Review* (Nairobi), 13 September 1991, 17.

184 "KANU—Maendeleo polls surprise outcome" *Weekly Review*, 10 November 1989.

185 "A contentious reckoning" *Weekly Review* (Nairobi), 1 December 1989, 14–15.
186 *Ibid.* Otete argued that MYWO had not received more than Ksh 10 million annually since 1952; at a later time, because she had been silenced during the meeting, she compiled her own counter-report with documentary evidence to show the "fictitiousness" of the Aburi report. It was Mrs Kiano, still patron of KMYWO, who silenced Otete, grabbing the microphone and forcing her to sit down.

Aburi made the following recommendations for KMYWO:

1 As the auditors for MYWO, Messrs Gichohi and company and Messrs Kimani and Onyancha and company "did not act responsibly" and should be dismissed.
2 Although KMYWO was an "independent" organization it should be overseen by KANU's Directorate of Youth and Women's Affairs. This Directorate would be responsible for coordinating KMYWO activities and controlling its foreign funds. This was a tremendous task KANU was willing to take on, as all of the monies in foreign financial assistance given to any and all women groups in Kenya would be overseen by this directorate.
3 KMYWO should dismiss its current top management staff for a permanent staff to be controlled by an executive director.

187 Macharia wa Mwati (1990) "Women's groups urged to link up" *Kenya Times*, 1 March 1990.
188 "Onsando assures women—Maendeleo to co-exist with other groups" *Daily Nation*, 16 February 1990.
189 "Women in search of self-reliance" *Daily Nation*, 14 November 1991.
190 Onyango 1992: 10.
191 Musa Jeffwa (1990) "Onsando lashes at foreign meddlers" *Kenya Times*, 30 October 1990.
192 *Ibid.*
193 "Saitoti is not a Maasai" (letter) in *Finance*, 15 March 1992, 10–11; "The struggle continues" *Drum*, April 1992, 6–9.
194 *Finance*, 16–31 December 1991, entire edition.
195 *Ibid*; "Taking the bull by the horns" in *Weekly Review*, 10 November 1989.
196 "The struggle continues" *Drum*, April 1992.
197 "Kenya aid is finally conditioned to reforms and democratization" *Finance*, 16–31 December 1991, 43–4; Jane Perez (1991) "Citing corruption by Kenyan officials, Western nations are cancelling aid" *New York Times*, 21 October 1991; *Standard*, 19 August 1991, 11 November 1991; *Nation*, 19 August 1991, 16 October 1991, 23 October 1991, 14 November 1991, 30 November 1991, 9 January 1992, 5 May 1992, 4 July 1992.
198 "Women support one-party system" *Standard*, 29 April 1990; "KMYWO condemns Odinga" *Kenya Times*, 14 May 1991.
199 *Ibid.*
200 "Shun evil-minded people says D.O." *Daily Nation*, 14 August 1990.
201 Humphrey Malalo (1990) "KMYWO pledge to back government," *Standard*, 30 October 1990.
202 "Women plan protest against pluralism" *Daily Nation*, 15 June 1990.
203 "Counter government critic—Aringo" *Daily Nation*, 13 March 1991.
204 Evans Lusveno (1991) "Women reject pluralism" *Kenya Times*, 29 May 1991.
205 *Ibid.*

206 Pius Nyamora (n.d.) "Wangari Maathai wrestling men" *Society*, 2/1, 10, 4–11; "That infernal Kenyan tower" *The Economist*, 3 February 1990, 41.
207 *Ibid.*
208 *Ibid.*
209 *Ibid.*
210 "Women flay Maathai" *Daily Nation*, 24 December 1989.
211 Nyamora (n. d.); Catherine Gicheru (1990) "Women to be disciplined" *Daily Nation*, 10 January 1990.
212 "Green-belt sign vandals to be punished—Onsando" *Standard*, 10 January 1990.
213 "Impressive record until now" *Weekly Review*, 21 February 1986.
214 Onyango 1992: 18.
215 Nzomo 1992b: 17.
216 "Wangari Maathai" *Presence*, 20.
217 "Taking the bull by the horns" *Weekly Review*, 10 November 1989.
218 Nzomo 1992b: 17.
219 Onyango 1992: 10.
220 "Women accuse chiefs" *Daily Nation*, 23 June 1990.
221 Wangui Gachie (1990) "Maendeleo owes Shs 1 million to KANU" in *Daily Nation*, 16 September 1990; Opala Kennethy (1990) "KMYWO owed Sh 2 million" *Daily Nation*, 16 September 1990.
222 "Women leader in for theft" *Daily Nation*, 28 September 1990; Waite Mwangi (1990) "Maendeleo official gets 3 years" *Standard*, 28 September 1990.
223 Patrick Wakhisi/Mung'ahu (1992) "Over 100 defect from KANU" *Standard*, 5 January 1992.
224 Kwendo Opanga (1991) "Aringo fired" *Daily Nation*, 23 December 1991.
225 Mwenda Njoka (1992b) "By-election fever" *Society*, 2, 5 (30 March 1992), 19–21; "Twists and turns" *Weekly Review*, 10 January 1992, 3–5; "Nation-wide elections to come" *Weekly Review*, 10 January 1992, 10–11; "The effect on Parliament of defections from KANU" *Daily Nation*, 7 January 1992.
Among those who created a crisis in Parliament by their defection were:

Name	*Constituency represented*	*Party defected to*
George Muhoho	Juja	Democratic Party (DP)
Mjenga Karum	Kiambaa	DP
John Gachui	Gatunga	DP
Kyale Mwendwa	Kitui West	DP
Oloo Aringo	Alego	Forum for the Restoration of Democracy (FORD)
Geoffrey Kare'ithi	Gichugu	FORD
Maina Wanjigi	Kamukunji	FORD
Kabibi Kinyanjui	Kikuyu	FORD
Mbooni	Mbooni	Social Democratic Party (SDP)
Peter Ejore	Turkana	Central

226 See the following for further discussions of Moi and KANU's opposition: "Moi time to go" edition, *Finance*, 16–31 December 1991; "Who will be the next President of Kenya?" *Finance*, 15 March 1992; "Is he [Matiba] fit to be president?" *Finance*, 15 April 1992.

227 Pauline Munene (1990) "Donors still helping body" *Kenya Times*, 10 January 1990.

228 "Women group has lost donor support" *Daily Nation*, 7 December 1991. Onyango 1992: 11.

229 *Ibid*; "Call to disengage organization, KANU" *Daily Nation*, 5 December 1991.

230 "Women's body, KANU affiliation severed" *Kenya Times*, 11 December 1991.

231 Onyango 1992: 18.

232 Pauline Munene (1991) "Aid not cut, says Onsando" *Kenya Times*, 11 December 1991.

233 "Hear our cries" *Society*, 2, 10 (4 May 1992), 14–15; 1991–2 interviews. All made references to the Wamwere incident and its impact on (K)MYWO projects.

234 1992 interview.

235 1991 interview.

236 This was obvious in the state's and MYWO's reaction to the Mothers' Hunger Strike of February–March 1992 and their joint continued harassment of Wangari Maathai. The mothers and supporters of 8 political prisoners, including Wamwere, who were being detained without trial, staged a peaceful hunger strike in Uhuru Park Freedom Corner commencing 28 February 1992. Maathai was one of the strikers. After 5 days of peaceful protest, on 3 March 1992, the Kenyan riot police savagely attacked the women with brute force, using tear gas and batons. Maathai was beaten unconscious and remained in a critical condition for 3 days. The mothers and grandmothers stripped half-naked, responding to police attacks and daring the police to kill them. They were cursing Moi and the cruelty and repression of his regime, and the police for being his henchmen. In many African traditions, stripping is a irrevocable curse of last resort.

KMYWO's chair, Onsando, and patron, Kiano, criticized the mothers for their stripping protest, calling their act of defiance "very shameful" and "unAfrican." In response to their criticism, Onsando and Kiano were challenged by women in Kenya to "tell who they represent." They were criticized for having "never raised a finger or come to the aid of their fellow women, yet they are supposed to be leaders."

See "Moi's day of shame" *Finance*, 15 March 1992, 4–5; Sheila Wambui, George Owour and Willys Otieno (1992) "Police break up 'freedom' demo" *Daily Nation*, 3 March 1992; "Starving for a just cause" *Standard*, 5 March 1992; "Beauty and the Beast: the curse of women fury" *Drum*, April 1992, 16–20. The author explains stripping as a curse in this way:

> Traditional folklore among various African tribes in Kenya and other countries speaks of what are known as curses and taboos in different societies. These curses and taboos have been with us from time immemorial and are strictly observed by all members of the society as failure to do so can result in the greatest punishment. The offender may be ostracized—made an outcast in his or her own community, die or run amok. . . . Stripping is a desperate act one takes as a last resort when a woman's anger has reached a boiling point. Historically the crux of the phenomenon was the imagery and symbolism applied to the protest against threats to women as a group or to an individual . . . the first act of women stripping took place 70 years ago . . . protecting the colonial regime's use of the kipande.

Drum April 1992: 16–20

"Maendeleo boss challenged" *Daily Nation*, 25 March 1992; "Wangari Maathai" *Presence*, January 1992.

237 Tom Matoke (1992) "Go for top seats, women advised" *Daily Nation*, 24 February 1992.

CHAPTER 4 A CHANGING RESEARCH METHODOLOGY AMID POLITICAL VOLATILITY, ENVIRONMENTAL UNCERTAINTY AND A CULTURE OF FEAR AND SILENCE

1 Kwendo Opanga (1991) "NGOs seek expulsion of human rights violators from C'wealth" *Nation*, 16 October 1991; "UK, Canada to punish human rights violators" *Nation*, 18 October 1991; "Bonn links aid to human rights" *Standard*, 11 November 1991.

2 "Kenya may lose US aid over corruption" *Daily Nation*, 23 October 1991; Jane Perez (1991) "Citing corruption by Kenya officials, Western nations are cancelling aid" *New York Times*, 21 October 1991.

3 Blamuel Njururi (1992) "Asians packing" *Society*, II, 9 (April 27), 37–8; "Citing corruption by Kenya officials, Western nations are cancelling aid" *New York Times*, 21 October 1991; anonymous interviews.

4 "Citing corruption by Kenya officials, Western nations are cancelling aid" *New York Times*, 21 October 1991.

5 Himbara 1994: 151–2; "Kenya announces economic reforms" *Reuter European Business Report*, 19 February 1993.

6 "Citing corruption by Kenya officials, Western nations are cancelling aid" *New York Times*, 21 October 1991; Muthui Mwai and KNA (1991) "Aid: I will not take this abuse, says Moi" *Daily Nation*, 19 September 1991; Makau Niko (1992) "IMF wants 2m fired—COTU boss" *Daily Nation*, 31 January 1992; Murigi Macharia (1991) "Stop blackmailing Africa, says Salim" *Kenya Times*, 18 November 1991; Argwings Odera (1992) "The hour of reckoning" *Society*, II, 5 (March 30), 45–6.

7 Blamuel Njururi (1992a) "Biwott: Ouko was my friend" *Society*, 20 April 1992, 28–30. This incident was discussed by many on the *magendo* networks in Kenya. Several journals and magazines have alluded to this incident in their writings. See *Weekly Review* issue 26 August 1994 and *Society* issue 30 March 1992. In East Africa, *magendo* refers to underground transactions, primarily economic ones. I am using *magendo* here to refer to an overall underground network of which connections and the economy are both part.

8 Argwings Odera (1992) "The hour of reckoning" *Society*, 2, 45–6; *Daily Nation* daily reports of Ouko Inquiry, 16 October 1990 to 26 November 1991; *Weekly Review* 23 August 1991, entire issue; Blamuel Njururi 1992a: 28–30; "Warped logic" *Weekly Review*, 11 October 1991, 16–20; anonymous interviews.

9 It was testified that Ouko had never taken commissions or received bribes.

10 "The Ouko Inquiry: the case of the 1988 election" *Weekly Review*, 23 August 1991, 6–7.

11 "Moi stops Ouko probe" and "Biwott, Oyugi, Orara arrested" *Daily Nation*, special edition, 26 November 1991; "Biwott, Oyugi and others arrested" *Daily Nation*, 27 November 1991; "The Ouko case: government issues a statement" *Weekly Review*, 10 January 1992, 17.

12 "*Society* magazine runs into trouble" *Weekly Review*, 10 January 1992, 16–17.

13 See note 236 to Chapter 3 for a discussion of the Mothers' Hunger Strike; Alexandra Tibbetts (1994) "Mamas fighting for freedom in Kenya" *Africa Today*, 41, 4, 27–48.

14 For information on LSK and the award, see "Law Society to get international award *Daily Nation*, 23 October 1991. For information on NCCK–government conflict, see Kenyan daily newspapers and weekly magazines from mid-1988 to 1989. For KPBWC's request for government to resign, see Njoroge wa Karuri (1992) "Govt asked to resign" *Daily Nation*, 18 May 1992. For government allegations against the press, see "*Society* magazine runs into trouble" *Weekly Review*, 10 January 1992.

15 Moi Day speech covered by all the daily newspapers, e.g. *Nation*, *Kenya Times*, *Standard*, 10 October 1991; Himbara 1994: 29.

16 "Donors: change in 6 months or else . . . " *Daily Nation*, 27 November 1991; Sinan Fisek (1991) "Politics dominated Kenya-donors meet" *Daily Nation*, 28 November 1991.

17 "KANU paves the way" *Kenya Times* special section: "KANU news a roundup of party activities" 7 December 1991.

18 "Ford launches party" *Daily Nation*, 7 December 1991.

19 Mwenda Njoka (1992a) "The killing fields" *Society*, II, 5 (30 March 1992).

20 "Chance to foster Kalenjin unity" *Weekly Review*, 13 September 1991), 12–14; Njoka 1992a; "Do the Kalenjins have a secret army?" *Drum*, April 1992, 28–9; Emman Omari (1992) "Biwott: I've no army in clash areas" *Daily Nation*, 20 March 1992; anonymous interviews. For Saitoti's alleged storage of arms, see daily newspapers for last half of the month of June 1992 and the first half of the month of July 1992.

21 "Kenya: multipartyism betrayed in Kenya—continuing rural violence and restrictions on freedom of speech and assembly" *Human Rights Watch/Africa*, 6, 5 (July 1994): 2–33.

22 In this clash, a young Luyhia man who was a school teacher had been murdered, allegedly by a group of Kalenjin, and his body had been dumped at a site of the Kenya National Cereals Board in Kipkarren. While I was there, Luyhia homes were being looted and burned. I was also delayed in Molo because of clashes (Molo borders Central and Rift Valley Provinces). In addition, I was cornered in my hotel room and questioned by a woman who I was told was a Kalenjin spy. She said that she was a Maasai and pretended to know me from a women's group meeting in an area in which I had never been. She prodded me about multipartyism, Moi, the clashes, the American Ambassador Hempstone—trying to get me to make seditious statements. I believe she taped the conversation. She prevented me from going to scheduled visits at various research sites in Kisii, Nyanza Province.

23 See Pearl Robinson (1994) "Democratization: understanding the relationship between regime change and the culture of politics", *African Studies Review*, 37, 1, 39–67 for a discussion of the culture of politics.

24 The Minister of Energy, Nicholas Biwott, was one of the key ministers shuffled by President Moi in September and October 1991 in a major government reorganization. In early 1992 Biwott was implicated and later arrested as a prime suspect in the murder of the former Minister of Foreign Affairs and International Cooperation, Dr Robert Ouko, in early 1992. In mid-1992 Biwott was further alleged to be a key instigator of "ethnic clashes," along with the Vice President George Saitoti.

25 KANU National Secretariat (n.d.) *Kanu: thirty great years, 1960–1990*, Nairobi: Government Printer, date not recorded.

26 *Ibid.*, 247–8. Babbie (1983) describes the various roles of observers in the field.

27 Yvonna Lincoln and Egon Guba (1985) *Naturalistic Inquiry*, Beverly Hills: Sage. The post positivist school is also identified by other names including ethnographic, subjective, qualitative, humanistic. The positivist school is also identified as objective, quantitative, scientific.

28 Babbie 1983; Lincoln and Guba 1985; David Dooley (1984) *Social Research Methods*, New Jersey: Prentice Hall.

29 As mentioned earlier, my first interview on the *magendo* initiated the snowball method of interviewee selection. That is, each interviewee referred me to other interviewees. So as to avoid getting only the *magendo* perspective, I also employed other methods of securing interviews. Some of the methods are relatively unorthodox, yet they were necessary given the research environment. For example, various procedures were used to contact women's groups. These procedures included: 1) going to the field and seeking groups out without previous contact; 2) meeting women leaders at community functions and arranging meeting times with the groups; 3) formally asking MYWO officials and staff to introduce me to groups; 4) establishing contact with women's groups through INGOs introductions; 5) writing letters and making phone calls to women leaders; 6) meeting women on the local *matatu* transportation system and establishing meeting times; 7) asking the district commissioner's office or the Peace Corps representatives to take me out into the field to meet groups; 8) going into the field with ministry personnel; and 9) sometimes asking the relatives of women's groups and friends to make introductions.

CHAPTER 6 THE WEB OF DECEIT: GRASSROOTS DEVELOPMENT CAUGHT?

1 For discussion of African feminism see Ama Ata Aidoo (1992) "The African woman today" *Dissent*, 39, 3 (Summer), 319–25; Filomena Steady (1981) *The black woman cross-culturally*, Massachusetts: Schenkman Publishing Company, specifically the introductory chapter; Rosalyn Terborg-Penn *et al. Women in Africa and the African diaspora*, 1–24, Washington DC: Howard University Press.

2 Fatton in Parpart and Staudt 1989: 49.

3 Bujra 1986: 135–7.

4 She was merely a figurehead as she was a very prominent woman in Kenya, in KANU and among women. Mjomba was a veteran women's leader, one of the few women recognized as KANU, and also the mayor of her municipality of Voi.

5 Bujra 1986: 135–7.

6 Psychological dependency on foreign donors to the extent that it leads to inactivity is discussed in a very interesting work by Nanda Shrestha "Becoming a development category" Jonathan Crush (ed.) (1995) *Power of Development*, 266–77, New York: Routledge.

7 Charlton 1984: 200.

8 *The United States Government Manual 1993/94*, Washington DC: United States Government Printing Office (1993), 707–10.

9 Wayne C. McWilliams and Harry Piotrowski (1993) *The World Since 1945*, 367–78, Boulder: Lynne Rienner Publishers.

10 1992 interview; Pathfinder International in Boston did not respond to repeated letters and phone calls I made to them to get more information.

11 1992 interview; Charlton 1984: 208.
12 *German Technical Agency Annual Report 1992*, Eschborn: Federal Republic of Germany, 1–9; telephone conversations with the Embassy of Germany in Washington DC.
13 Telephone conversations with Assistant Director of the KAF in Washington DC.
14 Telephone conversations with Cultural Counsel of the Embassy of Finland in Washington DC.
15 *The United States Government Manual 1993/94*, 815. USAID did not respond to questions in letters and phone calls in the US.
16 1992 interview; I also witnessed USAID "strong arm" Pathfinder while being disturbed in three separate meetings in progress which had to be canceled because of USAID demands.
17 1992 interview.
18 GTZ's agenda had developed as a response to a United Nations Conference on Energy Crisis in the 1970s. The SEP–Jiko project was aimed at reducing Kenya's spending on oil imports and focused on the efficient use of renewable energy resources in-country. GTZ collaborated with the Kenyan Ministry of Energy.
19 For a profound discussion, see Claude Ake (1994) *Democratization of Disempowerment in Africa*, 1–23, Lagos: Malthouse Press; Stephen Hellinger *et al. Aid for Just Development: Report on the Future of Foreign Assistance*, Boulder: Lynne Rienner Publishers.
20 With the quasi-government donors of Marttaliitto, GTZ and KAF–Marttaliitto offered advice and training to MYWO program managers when requested; and GTZ and KAF adopted a "hands-off" policy in the day-to-day implementation of the programs/projects they funded. With Coca-Cola, MYWO would be strictly monitored and penalized by Coca-Cola for not following established guidelines. MYWO's relations with USAID and the World Bank and the quasi-government donors they funded were all different, but they all required constant supervision, reporting, monitoring and meeting. It is important to note that only when Peace Corps volunteers received grants from USAID did the foregoing occur.
21 These sympathizers represent the pessimist school as presented in Chapter 2.
22 *Kenya Development Plan 1989–1993*, xix, 33–4.
23 Smith, Lethem and Thoolen discuss the larger environments within which organizations operate. Their arguments are presented in Chapter 2. De Graaf addresses an organization's environment in Zimbabwe. Twose looks at North–South partnerships.
24 Pringle's (1989) notion of women's counter-strategies of power alludes to this.
25 There were some ministry personnel who did complain of low rates and delays in pay. They tended to be civil servants with master's degrees who lived and worked in urban areas, and who had studied abroad.
26 The other reactions are 1) Monitoring and Registering, 2) Coordination and 3) Reorganization, Dissolution and Imprisonment. See Bratton 1989: 23–35.
27 These are discussed at length in Chapter 2.
28 Both the schools of thought which framed this research are presented at length in Chapter 2.
29 Patricia McFadden (1990) "Women in Southern Africa: five years after the decade" *African Commentary: A Journal of People of African Descent*, August, 30–2.
30 Hirono's perspective in IOR is presented in Chapter 2.

31 A hierarchy of donors was clear. USAID was at the apex, directing Pathfinder, CEDPA and the Peace Corps when volunteers received grants. The World Bank deferred to USAID for implementation. GTZ, KAF and Marttaliitto were regarded as a lower caliber of donors. Unlike USAID, they did not establish policy. Coca-Cola was also a lower caliber of donor. Of the donors, USAID and the World Bank were clearly the policy makers. The carried the most weight and received the most deference from all communities.

32 Nyoni 1987: 51; Bertrand Schneider (1988) *The Barefoot Revolution: A Report to the Club of Rome*, 230–1, London: Intermediate Technology Publications.

33 Schneider 1988: 230–1.

34 Garilao (1987) argues that Southern NGOs are more than mere conduits. MYWO defies his argument.

35 Johan Galtung (1980) *The True Worlds: A Transnational Perspective*, 133 and 310, New York: The Free Press.

36 Telephone conversation with David Cooperrider, Department of Interorganizational Behavior, Case Western Reserve University, Cleveland, Ohio, 18 June 1996.

37 This definition builds most specifically on Chad Alger (1987) "A grassroots approach to life in peace" *Bulletin of Peace Proposals*, 18, 3, and Chad Alger (1990) "Grassroots perspectives on global policies for development" *Journal of Peace Research*, 27, 2.

38 The complexity of the development enterprise globally is noted with caring skepticism by Marshall Wolfe (1996) *Elusive Development*, New Jersey: Zed Books. A strong call for a "moral imperative in development" is made by Dennis Goulet (1995) *Development Ethics: A Guide to Theory and Practice*, New Jersey, Zed Books.

39 Nzomo 1989: 12.

40 Esman and Uphoff 1984: 267.

BIBLIOGRAPHY

Adedeji, Adebayo (1993) "The case for remaking Africa" in *Action in Africa* edited by Douglas Rimmer, 53–7. New Hampshire: Heinemann.

African NGOs' Environment Network (ANEN) (1985) "The African crisis: where do we go from here?" Presentation at the United Nations Office of Emergency Operations for Africa/NGO Africa Emergency Consultation Meeting, Geneva, 22 July 1985.

Aidoo, Ama Ata (1992) "The African woman today" *Dissent* 39: 319–25.

Ake, Claude (1994) *Democratization of Disempowerment in Africa*. Lagos: Malthouse Press.

——(1996) *Democracy and Development in Africa*. Washington DC: The Brookings Institute.

Aldrich, Howard E. (1979) *Organizations and Environments*. New Jersey: Prentice Hall.

Aldrich, Howard E. and Whetten, David (1981) "Organization sets, action sets, and networks: making the most of simplicity" in *Handbook of Organizational Design* edited by Paul Nystrom and William Starbuck, 390. New York: Oxford University Press.

Alger, Chadwick (1978) "Extending responsible public participation in international affairs" *Exchange* 14: 17–21.

——(1979) "The organizational context of development: illuminating paths for wider participation." Contribution to the Sub-Project on Non-Territorial Actors of the GPID Project. Japan: The United Nations University.

——(1987) "A grassroots approach to life in peace" *Bulletin of Peace Proposals*, 18, 1–35.

——(1990) "Grassroots perspectives on global policies for development" *Journal of Peace Research* 27: 155–68.

Allison, Helen (1988) "Is foreign aid an asset?" *African Women Quarterly Development Journal* 1: 18–20.

Amin, Samir (1972) "Underdevelopment and dependence in black Africa: origins and contemporary forms" *Journal of Modern African Studies* 10: 503–24.

Ampoto, Akosua Adomako (1993) "Controlling and punishing women: violence against Ghanaian women" *Review of African Political Economy* 56: 103–11.

Anheier, Helmut K. (1987) "Indigenous voluntary associations, non-profits, and development in Africa" in *The Non-Profit Sector: A Research Handbook* edited by Walter W. Powell, 420–25. New Haven: Yale University Press.

Annis, Sheldon (1987) "Can small scale development be large scale policy?" *World Development* 15: 129–34.

Antrobus, Peggy (1987) "Funding for NGOs: issues and options" *World Development* 15 (supplement): 95–102.

Aubrey, Lisa (1995) "Women's differences in their perceptions and definitions of development: some reflections from Kenya" *Zumari* 2: 15–18.

——(1997a) "Toward a Pan African view of women and (Co) development: considering race in the era of development NGOs" *African Affairs Bulletin* 3.

——(1997b) "Who's leading development in Africa into the year 2000?: a look at the role of African and African American women in non-governmental organizations" paper presented at the 50th Anniversary Meeting of the 1945 Pan African Congress, University of Manchester, Manchester, England, October 1995: forthcoming 1997.

Babbie, Earl (1983) *The Practice of Social Research*. Belmont, California: Wadsworth Publishing Company.

Barkan, Joel, and Chege, Michael (1989) "Decentralizing the state: district focus and the politics of reallocation in Kenya" *Journal of Modern African Studies* 27 (3): 431–53.

"Beauty and the Beast: the curse of women fury" *Drum* (Nairobi), April 1992: 16–20.

Bebbington, Anthony, and Farrington, John (1993) "Governments, NGOs and agricultural development: perspectives on changing interorganizational relationships" *The Journal of Development Studies* 29: 199–219.

Bebbington, Anthony, and Thiele, Graham, with Penelope Davies, Martin Prager and Hernando Riveros (eds) (1993) *Non-Governmental Organizations and the State in Latin America: Rethinking Roles in Sustainable Agricultural Development*. New York: Routledge.

Bello, Walden (1994) *Dark Victory: The United States, Structure Adjustment, and Global Poverty*. London: Pluto Press.

Binder, L., Coleman, J.S., Pye, L.W., Weiner, M., LaPolombara, J., and Verba, S. (1971) *Crises and Sequences in Political Development*. New Jersey: Princeton University Press.

"Biwott, Oyugi and others arrested" *Daily Nation* (Nairobi), 27 November 1991: 1–2, 13.

Blau, Peter (1964) *Exchange and Power in Social Life*. New York: John Wiley.

"Bonn links aid to human rights" *Standard* (Nairobi), 11 November 1991.

Booth, David (ed.) (1994) *Rethinking Social Development: Theory, Research, and Practice*. Essex, England: Longman Scientific and Technical.

Boserup, Ester (1970) *Woman's Role in Economic Development*. New York: St. Martin's Press.

Boulding, Elise (1975) "Female alternatives to hierarchical systems, past and present" *International Associations* 6: 340–6.

Bratton, Michael (1989) "The politics of government—NGO relations in Africa" *World Development* 17: 569–87.

—— (1990) "Non-governmental organizations in Africa: can they influence public policy?" *Development and Change* 21, 87–118.

Brodhead, Tim (1987) "NGOs: in one year, out the other" *World Development* 15: 1–6.

Browne, Steven (1990) *Foreign Aid in Practice*. London: Printer.

Buijtenhuis, Robert (1971) *Mau Mau Twenty Years After: The Myth and Survivors*. Mouton: The Hague Press.

Bujra, Janet (1986) "Urging women to redouble their efforts . . . " in *Women and Class in Africa* edited by Claire Robertson and Iris Berger, 117–40. New York: Africana Publishing Company.

Burt, Ronald (1977) "Power in a social typology" *Social Science Research* 6: 1–83.
"Call to disengage organization, KANU" *Daily Nation* (Nairobi), 5 December 1991.
Cardoso, Fernando, and Faletto, Enzo (1979) *Dependency and Development in Latin America*. Los Angeles: University of California Press.
"Chance to foster Kalenjin unity" *Weekly Review* (Nairobi), 13 September 1991: 12–14.
Charlton, Sue Ellen M. (1984) *Women in Third World Development*. Boulder: Westview Press.
Chazan, Naomi (1989) "Gender perspectives on African states" in *Women and the State in Africa*, edited by Jane Parpart and Kathleen Staudt, 185–201. Boulder: Lynne Rienner Publishers.
Chazan, Naomi, and Rothchild, Donald (eds) (1988) *The Precarious Balance: State and Society in Africa*. Boulder: Westview Press.
Chazan, N., Mortimer, R., Ravenhill, J., and Rothchild, D. (1992) *Politics and Society in Contemporary Africa*. Boulder: Lynne Rienner Publishers.
Clark, John (1990) *Democratizing Development: The Role of Voluntary Organizations*. Hartford: Kumarian Press.
——(1995) "The State, popular participation, and the voluntary sector" *World Development* 23: 593–601.
"A contentious reckoning" *Weekly Review* (Nairobi), 1 December 1989: 14–15.
Cook, Karen S. (1977) "Exchange and power in networks of interorganizational relations" *The Sociological Quarterly* 18: 62–82.
Copestake, James G., and Welled, Kate (eds) (1993) *Non-Governmental Organizations and the State in Africa: Rethinking Roles in Sustainable Agricultural Development*. New York: Routledge.
"Counter government critic—Aringo" *Daily Nation* (Nairobi), 13 March 1991.
Cowen, Michael, and Shenton, Robert (1995) "The invention of development" in *Power of Development* edited by Jonathan Crush, 27–43. London: Routledge.
Crush, Jonathan (ed.) (1995) *Power of Development*. London: Routledge.
"Curb on use of vehicles" *Standard* (Nairobi), 21 December 1985.
Dahl, Robert (1957) "The concept of power" *Behavioral Science* 2: 201–15.
Darton, John (1994) "Survival test: can Africa rebound?" Special series of articles in *New York Times* 19, 20, 21 June 1994.
Davies, Celia (1992) "Gender, history and management style in nursing: towards a theoretical synthesis" in *Gender and Bureaucracy* edited by Mike Savage and Anne Witz, 229–52. Cambridge: Blackwell.
de Graaf, Martin (1987) "Context, constraint or control? Zimbabwean NGOs and their environment" *Development Policy Review* 5: 277–301.
Deutsch, Karl (1961) "Social mobilization and political development" *American Political Science Review* 55: 493–514.
Development Alternatives with Women for a New Era (DAWN) (1995) "Rethinking social development: DAWN's vision" *World Development* 23: 2001–4.
"Development NGOs" *Dossier* (Brussels), 104 (July–August 1987).
Deyo, Frederick (ed.) (1987) *The Political Economy of the New Asian Industrialism*. Ithaca: Cornell University Press.
Dictionary of International Relations Terms. Washington DC: Department of State Library (1987).
Dixon, William J (1976) "Research on research revisited: another half decade of quantitative and field research on international organizations" unpublished paper.
"Do the Kalenjins have a secret army?" *Drum* (Nairobi), April 1992: 28–9.

"Donors: change in 6 months or else . . . " *Daily Nation* (Nairobi), 27 November 1991.

Dooley, David (1984) *Social Research Methods*. New Jersey: Prentice Hall.

Dumont, Rene (1966) *False Start in Africa*. London: Sphere.

Edwards, Michael, and Hulme, David (1992) *Making a Difference: NGOs and Development in a Changing World*. London: Earthscan Publications.

"The effect on Parliament of defections from KANU" *Daily Nation* (Nairobi), 7 January 1992.

Efird, L. Julian (1977) "Test of bivariate hypotheses relating characteristics of international organizations to interorganizational relations" *The Emporia State Research Studies* 26, 5–52.

Ekeh, Peter (1975) "Colonialism and the two publics in Africa: a theoretical statement" *Comparative Studies in Society and History* 17 (1): 91–112.

Emerson, Richard (1962) "Power-dependence relations" *American Sociological Review* 27: 32.

Esman, Milton, and Uphoff, Norman (1984) *Local Organizations: Intermediaries in Rural Development*. Ithaca: Cornell University Press.

Evan, William (1976) *Organization Theory: Structures, Systems and Environments*. New York: John Wiley and Sons.

Evans, Peter (1979) *Dependent Development*. New Jersey: Princeton University Press.

"Excitement, protests mark Maendeleo polls" *Daily Nation* (Nairobi), 31 October 1989: 1, 13.

Farrington, John, and Lewis, David J., with S. Salish and Aurea Miclat-Teres (eds) (1993) *Non-Governmental Organizations and the State in Asia: Rethinking Roles in Sustainable Agricultural Development*. New York: Routledge.

Fatton, Robert (1989) "Gender, class, and State in Africa" in *Women and the State in Africa* edited by Jane Parpart and Kathleen Staudt, 48–9. Boulder: Lynne Rienner Publishers.

Feld, Werner (1979) *International Relations: A Transnational Approach*. California: Alfred Publishing.

Fernandez, Aloysius P. (1987) "NGOs in South Asia: people's participation and partnership" *World Development* 15 (supplement): 39–50.

Field, John, (ed.) (1993) *The Challenge of Famine*. Connecticut: Kumarian Press.

Fisek, Sinan (1991) "Politics dominated Kenya-donors meet" *Daily Nation* (Nairobi), 28 November 1991.

Fisher, Julie (in press) *Non-Governments: NGOs and Governments in the Third World*. Connecticut: Praeger.

"Ford launches party" *Daily Nation* (Nairobi), 7 December 1991: 1, 2.

Fosu, Augustin Kwasi (1992) "Political instability and economic growth: evidence from Sub-Saharan Africa" *Economic Development and Cultural Change* 40: 829–41.

Fowler, Alan (1985) "NGOs in Africa: naming them by what they are" In *Non-Government Organizations (NGOs): Contributions to Development: Proceedings of a Seminar Held at the Institute for Development Studies, Nairobi, 19 September 1985*, edited by Kabiru Kinyanjui, 7–30. Nairobi, Kenya: Institute for Development Studies.

——(1989) interview with author, Ford Foundation Kenya Office.

Frank, A. G. (1969) *Capitalism and Underdevelopment*. New York, Monthly Review Press.

Fregulia, Jeanette, and Okumu, Victoria (1984) untitled article about Jane Kiano, *Standard* (Nairobi), 29 April 1984.

Friedmann, John (1992) *Empowerment: The Politics of Alternative Development*. Cambridge, Massachusetts: Blackwell.

Gachie, Wangui (1990) "Maendeleo owes Shs 1 million to KANU" *Daily Nation* (Nairobi), 16 September 1990.

Galaşkiewicz, Joseph (1979) *Exchange Networks and Community Politics*. Beverly Hills: Sage.

——(1985) "Interorganizational relations" *Annual Review of Sociology* 11 : 281–304.

Galtung, Johan (1980) *The True Worlds: A Transnational Perspective*. New York: The Free Press.

Garilao, Ernesto D (1987) "Indigenous NGOs as strategic institutions: managing the relationship with Government and resource agencies" *World Development* 15: 113–20.

German Technical Agency (1992) *Annual Report*. Eschborn: Federal Republic of Germany.

Gezelius, Helena, and Millwood, David (1991) "NGOs in development and participation in practice" *Transnational Associations* 5: 278.

Gicheru, Catherine (1986) "Maendeleo boss gets the sack" *Daily Nation* (Nairobi), 14 February 1986.

——"Women to be disciplined" *Daily Nation* (Nairobi), 10 January 1990.

Godwin, Peter (1992) interview with author. World Bank staff in Kenya.

Gordenker, Leon, and Saunders, Paul (1978) "Organization theory and international organization" in *International Organization: A Conceptual Approach* edited by A.J.R. Groom and Paul Taylor. New York: Nichols Publishing.

Goulet, Denis (1995) *Development Ethics: A Guide to Theory and Practice*. New York: The Apex Press.

"Green-belt sign vandals to be punished—Onsando" *Standard* (Nairobi), 10 January 1990.

Groom, A. J. R., and Taylor, Paul (eds) (1978) *International Organization: A Conceptual Approach*. New York: Nichols Publishing.

A Guide to Women's Organizations and Agencies Serving Women in Kenya. Nairobi: Mazingira Institute (1985).

Hall, Richard (1991) *Organizations: Structures, Processes and Outcomes*. New Jersey: Prentice Hall.

Halonen, Anneli (1994) conversations with author. Cultural Counsel of the Embassy of Finland in Washington DC.

Hansen, Roger D. (1979) *Beyond the North–South Stalemate*. New York: McGraw Hill.

Harrison, Paul (1987) *The Greening of Africa*. Toronto: Paladin Grafton Books.

Hawley, Amos H. (1968) "Human ecology" in *International Encyclopedia of the Social Sciences* edited by D. L. Sills, 91–3. New York: Macmillan.

"Hear our cries" *Society* 2 (4 May 1992): 14–6.

Hellinger, S., Hellinger, D., and O'Regan, F.M. (1988) *Aid for Just Development: Report on the Future of Foreign Assistance*. Boulder: Lynne Rienner Publishers.

Himbara, David (1994) *Kenyan Capitalists, the State and Development*. Boulder: Lynne Rienner Publishers.

Hirono, Ryokichi (1987) "How to promote trilateral cooperation among governments, NGOs and international agencies" in *World NGO Symposium Nagoya Congress October 6–8, 1987, by the Organization for Industrial, Spiritual and Cultural Advancement International (OISCA)*. 77–85. Japan: OISCA.

hooks, bell (1981) *Ain't I a woman? Black women and feminism*. Boston: South End Press.

(1984) *Feminist Theory from Margin to Center*. Boston: South End Press.

Huntington, Samuel (1968) *Political Order in Changing Societies*. New Haven: Yale University Press.

Huxley, Elspeth (1968) *White Man's Country: Lord Delamere and the Making of Kenya*. New York: Praeger.

Hyden, Goran (1983) *No Shortcuts to Progress*. Berkeley: University of California Press.

"Impressive record until now" *Weekly Review*, 21 February 1986.

"Is he [Matiba] fit to be President?" *Finance* (Nairobi), 15 April 1992: 18–33.

Isaak, Robert (1995) *Managing World Economic Change: International Political Economy*. New Jersey: Prentice Hall.

Jeffwa, Musa (1990) "Onsando lashes at foreign meddlers" *Kenya Times*, 30 October 1990.

Jonsson, Christer (1986) "Interorganization theory and international organization" *International Studies Quarterly*, 30, 39–57.

Kahiga, Sam (1990) *Dedan Kimathi: The Real Story*. Nairobi: Longman.

Kajese, Kingston (1987) "An agenda for future tasks of NGOs: views from the South" *World Development* 15 (supplement): 79–86.

Kamau, Damaris (1985) "Report from the Thika Township Women Group" *Women's Voice* iii (3), 17.

Kanogo, Tabitha (1987) *Squatters and the Roots of Mau Mau 1905–1963*. Columbus: Ohio University Press.

"KANU and Maendeleo Ya Wanawake–a marriage full of promise" *Viva* (Nairobi), 5 (1988): 11.

"KANU bosses' wives win Maendeleo polls" *Standard* (Nairobi), 1 November 1989: 1, 12.

"KANU–Maendeleo elections surprise outcome" *Weekly Review* (Nairobi), 10 November 1989: 10.

KANU/MYWO (1987) Budget and Programme for proposed KANU Maendeleo Ya Wanawake Workshop, 18 March 1987.

KANU/MYWO Working Committee (1987) Minutes of Third Meeting on KANU/MYWO merger at KANU Headquarters, 3 March 1987, reference number MYWO/KANU/12/3; KANU/Maendeleo Ya Wanawake Organization Constitution and Rules, Objective Section, Number 4.

KANU National Secretariat (date not recorded) *Kanu: Thirty Great Years, 1960–1990*. Nairobi: Government Printer.

"KANU paves the way" *Kenya Times*, special section "KANU news: a roundup of party activities" 7 December 1991: 13–16.

Kanyinga, Karuti, and Ngethe, Njuguna (1992) "The politics of development space: the State and NGOs in the delivery of basic services in Kenya" Working Paper No. 486. Nairobi: Institute for Development Studies.

Karuri, Njoroge wa (1992) "Govt asked to resign" *Daily Nation* (Nairobi), 18 May 1992.

Kennethy, Opala (1990) "KMYWO owed Sh 2 million" *Daily Nation* (Nairobi), 16 September 1990.

"Kenya aid is finally conditioned to reforms and democratization" *Finance*, 16–31 December 1991: 43–4.

"Kenya announces economic reforms" *Reuter European Business Report*, 19 February 1993.

"Kenya may lose US aid over corruption" *Daily Nation* (Nairobi), 23 October 1991: 2.

Kenya Ministry for Culture and Social Services, Women's Bureau (1985) *Women of Kenya: Review and Evaluation of Progress*. Nairobi: Kenya Literature Bureau.

Kiano, Gikonyo (1991) conversation with author. Member of KANU party and government.

Kiano, Jane (1982) "What we stand for" *Viva* (Nairobi), 9 (special issue): 5–7, 73.

Kinyanjui, Kabiru (1989) "The African NGOs in context" presentation at NGOMESA Workshop on the theme of "The African NGO phenomenon: a reflection for action" Gaborone, Botswana, 15–19 May 1989.

Kipsongok, Joel (1989) "Don't be cheated, Moi tells women" *Standard* (Nairobi) 31 October 1989: 1–10.

"KMYWO condemns Odinga" *Kenya Times*, 14 May 1991.

Kobia, Sam (1985) "The old and the new NGOs: approaches to development" in *Non-Government Organizations (NGOs): Contributions to Development: Proceedings of a Seminar Held at the Institute for Development Studies, Nairobi, 19 September 1985* edited by Kabiru Kinyanjui, 35–6. Nairobi, Kenya: Institute for Development Studies.

Korten, David (1990) *Getting to the 21st Century: Voluntary Action and the Global Agenda*. Hartford: Kamarian Press.

Korten, David, and Brown, David C. (1988) "The role of voluntary organizations in development" an exploratory concept paper prepared for the World Bank, Institute for Development Research, Boston, Massachusetts, 27 September 1988.

Lasswell, Harold (1936) *Politics: Who Gets What, When, How.* New York: McGraw Hill.

Laumann, E. O., Galaskiewicz, J., and Marsden, P.V. (1978) "Community structure as interorganizational linkages" *Annual Review of Sociology 1978*, 458.

"Law Society to get international award" *Daily Nation* (Nairobi), 23 October 1991: 15.

Leftwich, Adrian (1996) *Democracy and Development*. Polity Press.

Legum, Colin (1953) "Mau Mau organization: psychological warfare needed" *The Scotsman*, 12 August 1953.

Leonard, David (1982) "Analyzing the organizational requirements for serving the rural poor" in *Institutions of Rural Development for the Poor* edited by David Leonard and D. R. Marshall, 1–38. Berkeley: Institute for International Studies.

Lerner, Daniel (1958) *The Passing of Traditional Societies: Modernizing the Middle East*. New York: The Free Press.

Levine, Sol, and White, Paul E. (1960–1) "Exchange as a conceptual framework for the study of interorganizational relations" *Administrative Science Quarterly* 5: 583.

Lewis, Shelby (1980) "African women and national development" in *Comparative Perspectives of Third World Women: The Impact of Race, Sex and Class* edited by Beverly Lindsay, 32. New York: Praeger.

——(1988) "A liberationist ideology: the intersection of race, sex, and class" in *Women's Rights, Feminism, and Politics in the United States* edited by Molly Shanley, 38–44. Washington DC: American Political Science Association.

Leys, Colin (1974) *Underdevelopment in Kenya: The Political Economy of Neocolonialism*. Los Angeles, University of California Press.

Liddle, R. William (1992) "The politics of development policy" in *World Development* 20 (6): 793–807.

Lincoln, Yvonna (1985) *Naturalistic Inquiry*. Beverly Hills: Sage.

Lincoln, Yvonna, and Guba, Egon (1981) "Criteria for assessing the trustworthiness of naturalistic inquiries" *Educational Communication and Technology Journal* 29 (2): 75–92.

Lissner, Jorgen (1977) *The Politics of Altruism—A Study of the Political Behavior of Voluntary Development Agencies*. Geneva: Lutheran World Federation.

Lusveno, Evans (1991) "Women reject pluralism" *Kenya Times* (Nairobi), 29 May 1991.

Macharia, Murigi (1991) "Stop blackmailing Africa, says Salim" *Kenya Times*, 18 November 1991: 1–2.

"Maendeleo fresh polls call" *Standard* (Nairobi), 17 June 1990.

"Maendeleo in financial mess" *Standard* (Nairobi), 10 November 1985.

"Maendeleo polls" *Daily Nation* (Nairobi), 19 September 1989.

"Maendeleo polls—a male affair?" *Weekly Review* (Nairobi), 3 November 1989: 4–7.

"Maendeleo to be probed" *Standard* (Nairobi), 11 November 1985.

Malalo, Humphrey (1990) "KMYWO pledge to back government" *Standard* (Nairobi), 30 October 1990.

"Male-dominated squabbling" *Weekly Review* (Nairobi), 27 October 1989: 14–5.

Markovits, Irving (ed.) (1987) *Studies in Power and Class in Africa*. New York, Oxford University Press.

Matoke, Tom (1992) "Go for top seats women advised" *Daily Nation* (Nairobi), 24 February 1992.

Mazrui, Ali A. (1986) *The Africans: A Triple Heritage*. London: BBC Publications.

Mbilinyi, Marjorie (1984) "Women in development ideology: the promotion of competition and exploitation" in *The African Review* 2 (1): 14–33.

Mbugua, Wanjiku (1989) "Women's body merges with ruling party" *Standard* (Nairobi), 10 October 1989: 29–30.

McWilliams, Wayne C., and Piotrowski, Harry (1993) *The World Since 1945*. Boulder: Lynne Rienner Publishers.

"Men meddle" *Standard* (Nairobi), 31 October 1989.

Metcalfe, L. (1981) "Designing precarious partnerships" in *Handbook of Organizational Design* edited by Paul Nystrom and William Starbuck, 505. New York: Oxford University Press.

Mitchell, J. C. (1969) "The concept and use of social networks" in *Social Networks in Urban Situations* edited by J.C. Mitchell, 2. Manchester: Manchester University Press.

Mohammed, Hussein (1991) "P. S. says Kenya has Sh 154 billion debt" *Kenya Times*, 19 November 1991: 10.

Moi, Daniel T. arap (1968) *Kenya African Nationalism: Nyayo Philosophy and Principles*. London: Macmillan.

"Moi stops Ouko probe" *Daily Nation* (Nairobi), 26 November 1991: 1–3.

"Moi time to go" *Finance* (Nairobi), 16–31 December 1991, entire edition.

"Moi's day of shame" *Finance* (Nairobi), 15 March 1992: 4–5.

Momsen, Janet Henshall (1991) *Women and Development in the Third World*. London: Routledge.

Moser, Caroline O.N. (1993) *Gender Planning and Development: Theory, Practice and* Training. New York: Routledge.

Mukhwana, Mayengo (1985) "Cooperate with the government" in *Women's Voice* 3: 1.

Mulford, Charles (1984) *Interorganizational Relations: Implications for Community Development*. New York: Human Sciences Press.

Munene, Pauline (1990) "Donors still helping body" *Kenya Times*, 10 January 1990.

——(1991) "Aid not cut, says Onsando" *Kenya Times*, 11 December 1991: 10.

Muthai, Wahome (1989) "Maendeleo elections postponed" *Kenya Times*, 29 August 1989.

——(1991) "Attitudes will change if women elites stand firm" *Daily Nation* (Nairobi), 23 September 1991.

"Mutisya tips on elections" *Daily Nation* (Nairobi), 18 September 1989.

Mwai, Muthui and KNA. "Aid: I will not take this abuse, says Moi" *Daily Nation* (Nairobi), 19 September 1991: 1–2.

Mwangi, Jacob (1986) "Towards an enabling environment: indigenous private development agencies in Africa" paper presented at the Enabling Environment Conference, Nairobi. Geneva: The Aga Khan Foundation.

Mwati, Macharia wa (1990) "Women's groups urged to link-up" *Kenya Times*, 1 March 1990.

Myrdal, Gunnar (1968) *Asian Drama: An Inquiry into the Poverty of Daily Nations*. New York: Pantheon.

"MYWO elections postponed once again" *Standard* (Nairobi), 22 September 1989: 32.

"Nationwide elections to come" *Weekly Review* (Nairobi), 10 January 1992: 10–11.

Ndegwa, Stephen N. (1996) *The Two Faces of Civil Society: NGOs and Politics in Africa*. Connecticut: Kumarian Press.

Ng'ethe, Njuguna (1991) *In Search of NGOs Towards a Funding Strategy to Create NGO Research Capacity in Eastern and Southern Africa*. Nairobi: Institute for Development Studies.

Niko, Makau (1992) "IMF Wants 2m Fired—COTU Boss" *Daily Nation* (Nairobi), 31 January 1992.

Njoka, Mwenda (1992b) "By-Election Fever" *Society* 2 (5): 19–21.

——(1992a) "The killing fields" *Society* 2 (5): 12–15.

Njururi, Blamuel (1992a) "Biwott: Ouko was my friend" *Society* 2 (8): 28–30.

——(1992b) "Asians packing" *Society* 2 (9): 37–8.

"Now Maendeleo top officer is suspended" *Standard* (Nairobi), 12 December 1985.

Nyamora, Pius (n.d.) "Wangari Maathai wrestling men" *Society* 2 (1): 4–11.

Nyang'oro, Julius (1994) "Reform politics and the democratization process in Africa" *African Studies Review* 37: 133–49.

Nyoni, Sithembiso (1987) "Indigenous NGOs: liberation, self-reliance, and development" *World Development* 15: 51–6.

Nzomo, Maria (1989) "The impact of the Women's Decade on policies, program and empowerment of women in Kenya" *Issue: A Journal of Opinion* 17: 9–17.

—— (1991a) "Women as men's voting tools" *Kenya Times* (Nairobi), 4 September 1991: 14–15.

——(1991b) "Women's passivity to blame for their woes" *Kenya Times*, 8 September 1991: 12, 29.

——(1992a) "Beyond structural adjustment programs: democracy, gender, equity, and development in Africa, with special reference to Kenya" in *Beyond Structural Adjustment in Africa: The Political Economy of Sustainable and Democratic Development*, 99–117. New York: Praeger.

——(1992b) "Schemes to divide and oppress women started in the colonial era" *The Sunday Daily Nation* (Nairobi), 17 May 1992: 17.

Odera, Argwings (1992) "The hour of reckoning" *Society* 2 (5): 45–6.

Odoyo, Odongo (1985) "MYWO looks to district focus" *Standard* (Nairobi), 12 November 1985.

Ogot, B.A. (1972) *Politics and Nationalism in Colonial Kenya*. Nairobi: East African Publishing House.

Ohene-Bekoe, K (1990) "Towards globalism in the 21st century: the human factor" *Texas Journal of Ideas, History, and Culture* 13: 4–7.

Okumu, Victoria (1983) "Oloitipitip advises women to raise cash" *Kenya Times*, 16 April 1983.

Omari, Emman (1989) "Maendeleo polls postponed again" *Daily Nation* (Nairobi), 22 September 1989: 1, 25.

——(1992) "Biwott: I've no army in clash areas" *Daily Nation* (Nairobi), 20 March 1992: 10.

"Onsando assures women—Maendeleo to co-exist with other groups" *Daily Nation* (Nairobi), 16 February 1990.
Onyango, Emily (1984) "Matiba, Kiano make up at last" *Kenya Times*, 6 June 1984.
Onyango, Otieno (1992) "The terminal crisis for Maendeleo" *Presence* (Nairobi), 8: 10–20.
Opanga, Kwendo (1991) "NGOs seek expulsion of human rights violators from C'wealth" *Daily Nation* (Nairobi), 16 October 1991.
Orora, John, and Spiegel, Hans (1980) "Harambee: self-help development project in Kenya" *International Journal of Comparative Sociology* 21: 243–53.
Osiemo, Nelson (1986) "Maendeleo boss sacked" *Standard* (Nairobi), 14 February 1986.
Otieno, Charles (1982a) "A history of Maendeleo" *Viva* (Nairobi), 9: 11–17, 88–90.
——(1982b) "Past chairmen" *Viva* (Nairobi), 9: 19–23, 85–7.
——(1982c) "An interview with Mrs Kiano" *Viva* (Nairobi), 9: 27–34.
——(1982d) "Maendeleo and the women of the world " *Viva* (Nairobi), 9: 37.
——(1982e) "At work in the Provinces" *Viva* (Nairobi), 9: 39–49.
——(1982f) "Committee members" *Viva* (Nairobi), 9: 63–71.
"The Ouko case: government issues a statement" *Weekly Review* (Nairobi), 10 January 1992: 17.
"The Ouko Inquiry: the case of the 1988 election" *Weekly Review* (Nairobi), 23 August 1991: 6–7.
Ouma, Sylvester J. (1980) *A History of the Cooperative Movement in Kenya 1908–1978*. Nairobi, Kenya: Bookwise.
Padron, Mario (1987) "Non-governmental development organizations: from development aid to development cooperation" *World Development* 15: 69–77.
Parpart, Jane (1995) "Post-modernism, gender, and development" in *Power of Development* edited by Jonathan Crush, 253–65. New York, Routledge.
Parpart, Jane, and Staudt, Kathleen (eds) (1990) *Women and the State in Africa*. Boulder: Lynne Rienner Publishers.
Parsons, Talcott (1960) *Structure and Process in Modern Societies*. Illinois: The Free Press.
Perez, Jane (1991) "Citing corruption by Kenya officials, Western nations are canceling aid" *New York Times*, 21 October 1991: A9.
Phillips, Anne (1991) *Engineered Democracy*. Pennsylvania: The Pennsylvania University Press.
"Politicians' wives in polls triumph" *Daily Nation* (Nairobi), 1 November 1989: 1, 6.
"Polls postponed" *Daily Nation* (Nairobi), 22 September 1989: 1, 25.
Potholm, Christian (1979) *The Theory and Practice of African Politics*. New Jersey: Prentice Hall.
Pradervand, Pierre (1988) "Self-reliance in Africa: peasant groups key to continent's survival" *Christian Science Monitor*, 9 March 1988, section 2: 1, 12.
——(1989) *Listening to Africa: Developing Africa From the Grassroots*. New York: Praeger.
"President's new rule on Maendeleo elections" *Daily Nation* (Nairobi), 27 October 1989.
Presley, Cora (1992) *Kikuyu Women, the Mau Mau Rebellion, and Social Change in Kenya*. Boulder: Westview Press.
Pringle, Rosemary (1989) *Secretaries Talk: Sexuality, Power and Work*. New York: Verso.

Public Record Office, Colonial Government of Kenya (1952a). *Community Development Organization—Kenya: Establishment and Annual Report for 1952* in file CO822/655, 1952–3, 237/13/01.

——(1952b) "Maendeleo Ya Wanawake: Extract from the Federation of Social Services Report for 1952" *Advancement of African Women in Kenya*, in file C0822/139, 1954–5, 269/5/04.

——(1952c) "Work in Districts" *Community Development Organization Annual Report for 1952.*

——(1953a) "Psychological warfare" Colonial House 1953, in file *Africa Information Services in Kenya.*

——(1953b) *Psychological Warfare in Kenya*. In file C0822/701, 1953, 389/313/01.

——(1953c) "Psychological warfare to all" *Psychological Warfare in Kenya* in file C0822/701, 1953, 389/313/01.

——(1953d) "Notes from the Working Party on Information" Provincial Commissioner, Central Province, 8 October 1953. *Psychological Warfare in Kenya*, file C0822/701, 1953, 389/313/01.

——(1953e) "Demand for women's clubs in Kiambu" Press Office Handout No. 840, 17 December 1953.

——(1954a) "Maendeleo Ya Wanawake: Extract from the Federation of Social Services Annual Report 1954" *Advancement of African Women in Kenya*, in file c0822/139, 1954–5, 269/5/04 .

——(1954b) "African women complete Domestic Science course" Press Office Handout No. 1044, 12 August 1954.

——(1954c) "European women build club for African women" Press Office Handout No. 1331, 22 October 1954. *Advancement of African Women in Kenya*, in file C0822/139, 1954–5, 269/5/04.

——(1954d) "Lady Baring Machakos safari" Press Office Handout No. 1351, 28 October 1954.

——(1954–5) "Women's clubs are "valuable rallying points" extract from Kenya Newsletter No. 85. *Advancement of African Women in Kenya*, in file C0822/139, 1954–5, 269/5/04.

——(1955a) "Progress report" Council of Ministers 1955.

——(1955b) "The Lady Mary Baring visits Kikuyu women's clubs" Press Office Handout No. 207, 24 February 1955.

——(1955c) "Women's club drive in Kiambu" Press Office Handout No. 621, 7 June 1955. *Advancement of African Women in Kenya*, in file C0822/139, 1954–5, 269/5/04.

——(1955d)."Classes for Kitale African women" Press Office Handout No. 965, 23 August 1955.

——(1955e) "Over 40,000 members of African women's clubs" Press Office Handout, 24 August 1955.

——(1956) "Terrorist strength" January 1956 estimates. *Future Emergency Policy in Kenya* in file C0008822/772, 1956, 15/02.

——(1957) *History of Mau Mau Emergency in Kenya*. In file C0822/220, 1957, 15/015.

——(1958) "Letter from Mrs. Prefumo to Miss Joan Vickers" 5 August 1958. *Maendeleo Ya Wanawake Clubs in Central Province*, in file C0822//665, 1958, 269/296/02.

——(n.d.) "African women's clubs" extract from *Kenya Newsletter* No. 90.

——(n.d.) "Women's clubs great progress in Machakos" Press Office Handout No. 47.

Rahman, M.D. Anisur (1993) *People's Self-Development: Perspectives on Participatory Action Research*. New Jersey: Zed Books.

Ramsay, Karen, and Parker, Martin (1992) "Gender, bureaucracy and organizational culture" in *Gender and Bureaucracy*, edited by Mike Savage and Anne Witz, 259–60. Cambridge: Blackwell.

Republic of Kenya (1973) "Sessional Paper No. 10 on Employment." Nairobi: Government Printer.

——(1989) *Development Plan 1989–1993*. Nairobi: Government Printer.

Rimmer, Douglas (ed.) (1993) *Action In Africa: The Experience of People Involved in Government, Business, and Aid*. London: Royal African Society.

Risah, Mumbi (1989) "Women's noble efforts to develop the nation" *Standard* (Nairobi), 10 October 1989.

Robertson, Claire (1984) *Sharing the Same Bowl: A Socioeconomic History of Women and Class in Accra, Ghana*. Bloomington: Indiana University Press.

Robinson, Pearl (1994) "Democratization: understanding the relationship between regime change and the culture of politics" *African Studies Review* 37: 39–67.

Rodney, Walter (1971) *How Europe Underdeveloped Africa*. Washington DC: Howard University Press.

Rogers, Barbara (1980) *The Domestication of Women: Discrimination in Developing Societies*. New York: St. Martin's Press.

Rosaldo, Michelle, and Lamphere, Louise (1974) *Women, Culture and Society*. Stanford: Stanford University Press.

Roseberg, Carl, and Nottingham, John (1966) *The Myth of "Mau Mau": Nationalism in Kenya*. New York: Frederick A. Praeger.

Rudestam, Kjell, and Newton, Rae (1992) *Surviving Your Dissertation*. Newbury Park: Sage.

"Saitoti is not a Maasai" letter to *Finance* (Nairobi), 15 March 1992: 10–11.

Sandbrook, Richard (1993) *The Politics of Africa's Economic Recovery*. Cambridge: Cambridge University Press.

Saul, John (1986) "The role of ideology in transition to socialism" in *Transition and Development: Problems of Third World Socialism* edited by R. Fagen, C.D. Deere, and J.L. Coraggio, 215. New York: Monthly Review Press.

Savage, Mike, and Witz, Anne (eds) (1992) *Gender and Bureaucracy*. Cambridge: Blackwell.

Schneider, Bertrand (1988) *The Barefoot Revolution: A Report to the Club of Rome*. London: Intermediate Technology Publications.

Scott, Catherine (1995) *Gender and Development: Rethinking Modernization and Dependency Theory*. Boulder: Lynne Rienner Publishers.

Sen, Gita and Grown, Caren (1987) *Development, Crises, and Alternative Visions*. New York: Monthly Review Press.

"Sh 40 million for KMYWO projects" *Daily Nation* (Nairobi), 2 July 1990.

Shrestha, Nanda (1995) "Becoming a development category" in *Power of Development*, edited by Jonathan Crush, 266–77. New York: Routledge.

Smillie, Ian (1995) *The Alms Bazaar: Altruism Under Fire Non-Profit Organizations and International Development* Ottawa: International Development Research Center.

Smith, Brian H. (1990) *More Than Altruism: The Politics of Private Foreign Aid*. New Jersey: Princeton University Press.

Smith, Joan, Wallerstein, I., and Evers, H.-D. (eds) (1984) *Households and the World Economy*. Beverly Hills: Sage.

Smith, W.E., Lethem, F.J., and Thoolen, B.A. (1980) "The design of organizations for rural development projects: a progress report" *World Bank Staff Working Papers* No. 375. Washington DC: World Bank.

"*Society* magazine runs into trouble" *Weekly Review* (Nairobi), 10 January 1992: 16–17.

"Starving for a just cause" *Standard* (Nairobi), 5 March 1992: 16.

Staudt, Kathleen (1980) "Women's organizations in rural development" report distributed by the Office of Women in Development, US Agency for International Development. Washington DC: USAID.

Staudt, Kathleen, and Glickman, Harvey (1989) "Beyond Nairobi: women's politics and policies in Africa revisited" *Issue: A Journal of Opinion* XVII (2): 4–6.

Steady, Filomena (1981) *The Black Woman Cross-Culturally*. Massachusetts: Schenkman Publishing.

Steel, David J. (1987) "Commemorative lecture" in *World NGO Symposium Nagoya Congress October 6–8, 1987, by the Organization for Industrial, Spiritual and Cultural Advancement International (OISCA)*, 139. Japan: OISCA.

Steifel, Mathias, and Wolfe, Marshall (1994) *A Voice for the Excluded: Popular Participation in Development—Utopia or Necessity?* New Jersey: Zed Books.

Steihm, Judith H. (ed.) (1984) *Women's Views of the Political World of Men*. New York: Transnational Publishers.

Steintra, Deborah (1994) *Women's Movements and International Organizations*. New York: St. Martin's Press.

Stoltenberg, John (1977) "Toward gender justice" in *For Men Against Sexism* edited by Jon Snodgrass, 75–6. California: Times Change Press.

"Taking the bull by the horns" *Weekly Review* (Nairobi), 13 September 1991: 17.

"The struggle continues" *Drum* (Nairobi), April 1992: 6–9.

Tandon, Yash (1990) "Foreign NGOs, uses and abuses: an African perspective" excerpt from a paper read at the Roundtable organized by the African Association of Public Administration (AAPAM) and the United Nations Economic Commission for Africa Special Action Program in Administration and Management—Regional Project (ECA/SAPAM), on the theme "Mobilizing the informal sector and non-governmental organization for recovery and development: policy and management issues, Abuja, Nigeria, December 1990.

——(1978) "The interpretations of international institutions from a Third World perspective" in *International Organization: A Conceptual Approach* edited by Paul Taylor and A. J. R. Groom, 357–38. New York: Nichols Publishing.

Taylor, Debbie (ed.) (1985) *Women: A World Report*. New York: Oxford University Press.

Tendler, Judith (1975) *Inside Foreign Aid*. Baltimore: John Hopkins University Press.

——(1982) *Turning Private Voluntary Organizations Into Development Agencies*. Washington DC: US Agency for International Development.

Terborg-Penn, Rosalyn, Harley, Sharon, and Benton-Rushing, Sharon (eds) (1987) *Women in Africa and the African Diaspora*. Washington DC: Howard University Press.

"That infernal Kenyan Tower" *The Economist*, 3 February 1990: 41.

Tibbetts, Alexandra (1994) "Mamas fighting for freedom in Kenya" *Africa Today* Fourth Quarter 1994: 27–48.

"To the women we say: vote wisely" *Daily Nation* (Nairobi), 30 October 1989.

Topouzis, Daphne (1990) "The feminization of poverty " *Africa Report* July/August 1990: 60–3.

"Twists and turns" *Weekly Review* (Nairobi), 10 January 1992: 1–4.

Twose, Nigel (1987) "European NGOs: growth or partnership?" *World Development* 15 (supplement): 7–10.

"UK, Canada to punish human rights violators" *Daily Nation* (Nairobi), 18 October 1991.

Union of International Associations (ed.) (1993–4) *Yearbook of International Organizations*, Appendix 2. London: Agreement with the United Nations.

United Nations (1985) *Statement by the Non-Governmental Organizations Attending the Eleventh Ministerial Session of The World Food Council*. Paris, 10–13 June 1985. New York: United Nations.

United Nations (1994) *The United Nations Development Program 1994 Human Development Report*. New York: United Nations.

United States of America (1993) *United States Government Manual 1993/94*. Washington DC: United States Government Printing Office.

"The University of Nairobi: a history of intimidation" *Finance*, 16–31 December 1991: 24–7.

Van de Ven, Andrew, and Ferry, Diane L. (1980) *Measuring and Assessing Organizations*. New York: John Wiley.

van der Heijden, Hendrik (1987) "The reconciliation of NGO autonomy, program integrity and operational effectiveness with accountability to donors" *World Development* 15 (supplement): 103–12.

——(1990) "Efforts and programmes of NGOs in the least developed countries" *Transnational Associations* 1: 19.

Wakhisi, Patrick (1991) "Maendeleo branch members want polls cancelled" *Standard* (Nairobi), 1 February 1991.

Wambui, Sheila, Owour, George, and Otieno, Willys (1992) "Police break up 'freedom' demo" *Daily Nation* (Nairobi), 3 March 1992: 1, 3.

Wandal, Haroun, and bureaux reporters (1989) "Men meddle in Maendeleo polls" *Standard* (Nairobi), 31 October 1989: 1, 3.

"Wangari Maathai" *Presence* 8 (1992): 3–6, 12–14, 20, 30.

Wanyeki, Grace N. (1985) "Report from Central Province" *Women's Voice: Official Journal of Maendeleo Ya Wanawake* lll (3): 16.

"Warped logic" *Weekly Review* (Nairobi), 11 October 1991: 16–20.

Whetten, David A. (1981) "Interorganizational relations: a review of the field" *Journal of Higher Education* 52 (1): 1.

Whitaker, Jennifer Seymour (1988) *How Can Africa Survive?* New York: Council on Foreign Relations Press.

"Who will be the next President of Kenya?" *Finance* (Nairobi), 15 March 1992: 20–38.

Wipper, Audrey (1971) "Equal rights for women in Kenya?" *The Journal of Modern African Studies* 9 (3): 429–42.

——(1972) "African women, fashion, and scapegoating" *Canadian Journal of African Studies* VI (ii: 329–49.

——(1975) "The Maendeleo Ya Wanawake Organization: the co-optation of leadership" *African Studies Review* XVIII (3): 99–120.

——(1975–6) "The Maendeleo Ya Wanawake movement in the colonial period: the Canadian connection, Mau Mau, embroidery and agriculture" *Rural Africana* 29: 195–214.

Wolfe, Marshall (1996) *Elusive Development*. New Jersey: Zed Books.

"Women accuse chiefs" *Daily Nation* (Nairobi), 23 June 1990.

"Women body faces legal suits" *Standard* (Nairobi), 6 February 1986.

"Women's body, KANU affiliation severed" *Kenya Times*, 11 December 1991: 4.

"Women flay Maathai" *Daily Nation* (Nairobi), 24 December 1989: 5.

"Women group has lost donor support" *Daily Nation* (Nairobi), 7 December 1991.

"Women in search of self-reliance" *Daily Nation* (Nairobi), 14 November 1991.

"Women plan protest against pluralism" *Daily Nation* (Nairobi),15 June 1990.
"Women support one-party system" *Standard* (Nairobi), 29 April 1990.
Woman's Voice: Official Journal of The Maendeleo Ya Wanawake III (1), (April 1985).
Woman's Voice: The Official Journal of The Maendeleo Ya Wanawake III (2), (May 1985).
World Bank (1994) *Adjustment in Africa: Reforms, Results, and the Road Ahead.* New York: Oxford University Press.
Yudelman, Sally (1987) "The integration of women into development projects: observations on the NGO experience in general and in Latin America in particular" *World Development* 15 (supplement): 179–87.

INDEX

Aburi, A. M. 79
accountability 81; *see also* Maendeleo Ya Wanawake
African Development Bank 82
African development policies 2
African NGOs 2, 24–30, 99, 160, 166
African Studies 15
Africanization 52, 54
Africare 23
Agina, Mareso 77
aid 141, 146; bilateral 21, 151; British 26; foreign 15, 91, 129; multilateral 21; private 21
Ake, Claude 1
Aldrich, Howard 14, 43
Alger, Chadwick 11, 12
Allison, Helen 26
The Alms Bazaar 10
Alouch, Phoebe 77
Amayo, David 73
American Bar Association 95
Amin, Samir 31
Ampoto, Akosua Adomako 34
Anguka, Jonas 94
Antrobus, Peggy 25
Aringo, Peter oloo 74, 83, 85
aristocracy 48
Asia 15, 31
Asians *see* ethnicity
Asiyo, Phoebe 52, 54–5
Askwith, T. G. 48
Associated Country Women of the World (ACWW) 50, 53
authoritarianism 81
autocracy 82, 90
autonomy: definition 173; *see also* MYWO; NGO

Baring, Mary 48–9
Baringo 49
Bebbington, Anthony 15
Binder, Leonard 31
Binns, H. K. 46
Biwott, Nicholas 93, 94, 97
Blau, Peter 41, 158
Booth, David 33
Boserup, Ester 29
Boulding, Elise 36
Bratton, Michael 27
Bretton Woods Conference 150
Brodhead, Tim 192n86
Brown, L. David 28
Brussels 21
Bujra, Janet 37, 38
bureaucrats *see* civil service; research strategy
Burkino Faso 25
Burt, Ronald 39, 158
businesses: Kenyan 23–4, 153; transnational 23–4, 150

Canada 82
capitalism 127, 128
censorship *see* Kenyan government
Center for Population and Development Activities (CEDPA) 101, 125, 127, 133, 150, 151
Central Organization of Trade Unions of Kenya (COTU) 67, 68, 139
Central Province 47, 48, 76
Charlton, Sue Ellen 30
Chazan, Naomi 36
Chinese Womens Delegation 104
Christian Aid 27
Christian Health Association of Kenya (CHAK) 110

civil service 24, 102, 132–8, 156–8, 161
civil society 14, 15; civic public 32; culture of fear and silence 89–98, 107, 132; culture of politics 32, 98, 99; primordial public 32
class *see* Maendeleo Ya Wanawake
Coast Province 69
Coca Cola 24, 60, 127, 128, 150, 151
Coca Cola–Kenya 23, 60, 101, 153
Colonialism: Community Development Annual Report of 1952 49; concentration camps 47; Department of Community Development 46–50; forced labor 48; government 46–57,138; Kamba home guard 49; Local Native Councils 55; Ministry for Community Development and Rehabilitation 48; missionaries 46; national resistance 46–57; press office 50, 51; pschological warfare 47, 51, 52; settlers 52; wives 46, 52; womens groups 46–50
Commission of European Communities 82
Commonwealth Countries 90
comprador 31, 165
Cook, Karen 158
Copestake, James 15
The Courier 21
culture: cultural imperialism 20; culture of fear and silence 89–98, 107, 132; culture of politics 32, 98, 99

Dahl, Robert 41
Daily Nation 74, 103, 129, 140
Dandora Ward 75
Danish International Development Agency (DANIDA) 103, 110, 133, 141
Dar, Mrs 59, 70
David, Mary 49
Davies, Celia 36
de Graaf, Martin 17–19, 26, 164
Decade for Women *see* United Nations Decade for Women; women
democracy 127, 128, 153
Democratic Movement (DEMO) 96; *see also* multipartyism
Democratic Party (DP) 96; *see also* multipartyism
Democratization 92; *see also* multipartyism
Denmark 82, 91
dependence *see* Maendeleo Ya Wanawake; RDM; women
development: agenda 1, 2, 163, 164; anti-development 161, 165; community development 12; control 16; cooperation 9; cooperative endeavours 1, 4, 5; dependency theory 30–1; definition 165, 166; ethics 17; enterprises 23; entrepreneurship 128; failures 2, 163; funding 5, 61; interorganizational cooperation 9, 10; male bias 30–4, 37; models 31; modernist framework 1, 33; modernization theory 30; participation 165; partnerships 1, 4, 5; scholarship 2, 8, 31; strategy 64–7; success 2, 166 ; theory 7, 8, 30–4; underdevelopment 30, 31; world systems theory 31
Development Alternatives with Women for a New Era (DAWN) 194n116
development cooperation *see* development
development partnerships 122; African governments 2–6, 20–29, 54–88; attitudes 16, 17, 71, 72, 73–83, 85–8; bureaucracy 9, 60, 64, 65, 69, 86; businesses 23, 24, 59, 60, 83; control 16; cooperation 143; decision-making power 71, 203n144; domination 39; ethics 17; financial dependency 18–19; focal organization 41–3, 127, 131, 147; influence 70–72, 203n144, 86; interorganizational cooperative endeavors 1, 9, 10, 20–4, 38, 45, 50, 51, 53, 56–60, 62, 63, 66, 68, 70, 78, 81, 83, 86–8, 112; linkages 9, 53, 56–60, 62, 63, 66, 70, 88; mutual hostility 16; optimist school 5, 6, 25, 26, 160, 167; pessimist school 5, 6, 26–9, 160, 167; power 38–41, 129; stategic interdependence 42; suspicion 16–17, 132; trilateral partnerships 112, 116, 117, 122–4, 125, 129, 130–2, 135, 136, 146–9, 152, 156, 157, 160, 161, 163, 164, 167, 168;

Web of Deceit 160–8; women's organizations 36–44
development policy process 131, 140; evaluation 120, 131; formulation 121, 128, 129, 145; implementation 121, 122, 128, 129, 130, 134, 145, 149, 151, 162; political actors 140, 141
Dolphin Club 91

East Wanga Location 75
Eastern Province 49
Eastleigh 60
economic development *see* Kenya
economic liberalization *see* structural adjustment programs
economic policy *see* Kenyan government
economy *see* Kenya
education *see* women
Efird, L. Julian 10
elections *see* MYWO 1989 elections
Elliot, Charles 27
Embu 49
Emerson, Richard 39, 41
equality *see* gender, Maendeleo Ya Wanawake; women
Esman, Milton 28, 166
ethics *see* development; development partnerships
Ethiopian famine 2
ethnicity: clashes 114; ethnocentrism 97; European 45, 52; George Saitoti 81; Gikuyu Embu Meru Association (GEMA) 55, 58; Indians (Asians) 59, 60, 91; Jomo Kenyatta 55, 58; Kalenjin 91, 96–7; Kikuyu 55, 96; Kisii 96; Luo 96; Luhyia 96; Maasai 96
European Investment Bank 82
Evan, William 41, 42

Farrington, John 15
Fatton, Robert 34
feminism 107, 145, 164; anti-feminists 144; feminist theory 167; feminists 145
Ferry, Diane 14
fieldwork *see* research strategy
Finance Magazine 82
financial assistance 98; *see also* foreign donor assistance; Kenyan government; Maendeleo Ya Wanawake
Finland 82, 151
Finnish government 71, 130
Ford Foundation 60, 71, 130
Foreign Assistance Act 150
foreign donors: 22, 71, 90, 93, 98, 101, 106, 127–37, 139, 149; Consultative Group 82; donor community 101, 102, 108, 129, 140; hegemony 116; humanitarianism 127, 128, 130; imperialism 164; influence 11, 40, 149–57; linkages to home governments 151–4; withdrawals 86, 87, 90, 92
foreign donor assistance: financial 101, 112; motivational 111; ongoing 112; push/crisis 112; technical 101, 112
Fort Hall 49
Fort Jesus 83
Fowler, Alan 20
France 82

Galaskiewicz, Joseph 13
Galtung, Johan 164
Gamble, Clarence 180
Garilao, Ernesto 15–16
Garissa 49
gender 147; equality 34, 36, 147; Gender and Development (GAD) 166; identity 34–5; inequalities 34, 36, 147; male bias *see* development; male domination 17; male egos 59, 164; misogyny 97; patriarchy 147; planning 34; private sphere 33, 147; public sphere 30, 33, 147; relations 36
General China 48
general systems theory 41; input organizations 41; output organizations 41
German Agency for Technical Cooperation (GTZ) 22, 62, 101, 127, 128, 130, 132, 133, 150, 151, 152, 153
Germany 81, 82, 90
Ghana 2
Githeka, Nelia 77
Glickman, Harvey 37
global interdependence 12, 166
global North *see* North
global South *see* South

Gordenker, Leon 9, 12, 13
governmental organizations (GOs) 22
grassroots *see* women
grassroots sector organization (GROs) 23
Green Belt Movement (GBM) 23, 28, 83, 84, 119; *see also* Maathai, Wangari

Habwe, Ruth 56
harambee 58, 64–6, 115; *see also* Kenyan government; Maendeleo Ya Wanawake
Hawley, Amos 13
Hempstone, Smith 91
Hirono, Ryokichi 16–19
hooks, bell 34, 35
House of Manji 59–60
human rights 81, 84, 113; violations 7, 81, 90, 91, 125
Huntington, Samuel 31

Ichima, Haroun 75
Indians *see* ethnicity
indigenization 52
interdependence *see* development partnerships; global interdependence; inter-organizational relations
International Development Cooperation Agency (IDCA) 151
international governmental organizations (IGOs) 22
International Monetary Fund (IMF) 2, 92, 99, 140, 154, 158, 163, 164
international nongovernmental organizations (INGOs) 22
interorganizational relations (IOR) 10–19, 89, 146; balance of power 40–89; focal organization 41–3, 127, 131, 147; influence 168; input organizations 41–2; interdependence 147; linkages 9–15, 38, 39, 41–4, 45, 53, 56–60, 62, 63, 66, 88; linking pin 42, 43, 147; network analysis 13–15, 41–4, 156; network linkages 9, 13–15, 45, 147; networks 13–15; output organizations 41–2; power 10, 15, 16, 34–44; relational environment 13–15; strategic interdependence 42; theory 9–15; transnational interorganizational linkages 10–15, 38, 39, 41–4, 45–51, 53, 56–60, 62, 63, 66, 67, 78, 81–3, 86–8, 164; women's organizations 38–44, 113
international relations 11–13, 166, 168
Italy 82
Iveti's Women's Club 49

Japan 82
Japanese International Cooperation Agency (JICA) 141
jiko *see* MYWO national programs
Jonsson, Christer 13, 42–3

Kakamega 85
Kalenjin 91, 97; *see also* ethnicity
Kanogo, Tabitha 48
Kasarani 96
Keen, John 86
Kenya: economic conditions 68; economic development 64–8, 96; economy 96; multiparty politics 85, 86; National Development Plans 65, 140; national development policies 64–8; political instability 89–98; recolonialization 3, 5; State of Emergency 46
Kenya Africa National Union (KANU) 90–2; affiliation of MYWO 4, 7, 70–2, 73–80, 86; appendage 7, 68; benefactor 72; beneficiary 72; defections 85, 86, 207n225; disaffiliation 4, 86–8; Directorate of Youth and Womens Affairs 73, 79, 109; hegemony123; Patriarchy 124; Womens League 109, 144; Womens Wing 74; Youth Wingers 76
KANU Maendeleo Ya Wanawake ((K)MYWO) 45, 70–88, 110, 130
Kenyan Africa Democratic Union (KADU) 90
Kenya Institute of Administration 46, 72
Kenya Lion's Club *see* Maendeleo Ya Wanawake
Kenya National Archives 103
Kenya Professional and Business Womens Club (KPBWC) 95, 142
Kenya Times 103
Kenya Times Media Trust (KTMT) 83–5 *see also* Maathai, Wangari
Kenyan government: assistance to MYWO 4, 24, 38, 43, 44;

bureaucrats 132–8, 156, 157, 161; censorship 95; Central Bureau of Statistics 74; Civil Service 92, 132, 150, 158; civil servants 132–8; Commissioner of Income Tax 69; constitution 90, 96; Constitutional changes 90, 96; corruption 78, 80–1, 90–3, 96, 123, 142; culture of fear and silence 98; culture of politics 98; debt 68; economic development policy 64–8; financial assistance to MYWO 4, 43–4, 56–8, 70, 82, 112; harambee 58, 64–6, 115; hegemony 123; influence on MYWO 63–72, 79–81, 85–8, 203n144; Ministry of Agriculture 24, 102, 132, 133, 136, 137; Ministry of Agriculture Home Economics Extension 102, 136; Ministry of Culture and Social Services (MCSS) 24, 60, 64, 69, 70, 86, 102, 113, 136, 138–41; Ministry of Energy 24, 102, 133; Ministry of Foreign Affairs and International Cooperation 93, 94 ; Ministry of Health 24, 102, 113, 133; Ministry of Home Affairs and National Heritage 24, 102, 132, 136; Ministry of Home Affairs and National Heritage's National Council for Population and Development (NCPD) 102, 123, 133, 137; Ministry of Livestock Development 24, 102, 132, 138; Multiparty state 81, 85–6; Nyayo philosophy 66, 81–2, 86; one party state 80, 81; Parliament 207n225, 67, 86; patriarchy 55, 69, 71; personnel 132–8, 156; political climate 89–98; political instability 89–98; political and economic reform 90–98; repression 132, 142; sovereignty 129, 154, 162; spies 98; transparency 113; technical assistance to MYWO 4, 43–4, 112; Women's Bureau 65, 130, 139
Kenyan government opposition 85, 86, 90; citizens 7; external forces 7, 90–2; internal forces 7, 93–8; international community 7, 90–2, 99; Mother's Hunger Strike 95, 208n236; students 82
Kenya Peoples Union 90
Kenyatta, Jomo 54, 55, 58, 60, 64, 65, 82, 90
Kericho 50
Kiamaa 75
Kiambu 49
Kiano, Jane 57, 70, 78, 79, 84, 87, 88, 99, 105, 107, 115–17, 121, 128, 143, 137, 139, 143
Kiano, Julius Gikonyo 56–63
Kibaki, Mwai 86
Kikuyu *see* ethnicity
Kilifi 50, 84
Kimathi, Dedan 48
Kirinyaga 82
Kirui, Alice arap 50
Kirui, Jane 68, 77, 85
Kisii 63
Kisumu Ndogo Ward 75
Kitale 50
Kitui 49
Kinyanjui, Kabiru 26, 27, 154
Kobia, Sam 26, 27, 95, 154
Kodia, J. A. 76
Konrad Adenauer Foundation (KAF) 58, 62, 81, 86, 101, 126–8, 130, 131, 150–3
Korten, David C. 28

Lady Grey 48
Lamu 56
Langata 83
Latin America 15
Laumann, Edward 13, 14
Law Society of Kenya 67, 68, 83, 95; International Human Rights Award 95
League of American Women Voters 53
Lecomte, Bernard 25
Lenana Ward 75
Leonard, David 9
Lethem, Francis J. 17
Levine, Sol 39, 41
Lewis, David 15
Lewis, Shelby 31, 53
Leys, Colin 31
liberalism 1, 128, 153
Liddle, R. William 32
Lindsay, Beverly 31
Lloyd, Frank 48
Lodwar 54
Luo *see* ethnicity
Luhyia *see* ethnicity

Maasai *see* ethnicity
Maasai Ward 75
Maathai, Mwangi 83,
Maathai, Wangari 68, 83, 84, 94, 119, 137; Green Belt Movement (GBM) 83–5; Kenya Times Media Trust 92; Uhuru Park 119
Machakos 49, 74
Maendeleo Ya Wanawake (MYWO): affiliation 4, 7, 63–8, 70–2, 73, 119, 124, 130, 132, 138, 139, 144, 155, 157, 203n144; agenda 129, 139, 140, 154, 163, 164; anti-feminists 114; appendage 7, 68, 72, 162; autonomy 4, 38, 45, 55, 53, 55, 71, 80, 87, 89, 98–100, 119, 124, 130, 132, 135, 149, 154, 156, 160, 163; benefactor 72; beneficiary 72; caretaker committee 70, 202n143; childcare programs 56; class 138, 157, 165; community based distributors 126; community development 47; conservative ideology 58, 59, 66; constitution 108; constitutional changes 71, 203n144; cooptation 67–72; corruption 69, 141; coup 52–3; creditors 70, 201n136, 202n143; decision-making power 71, 116, 203n144; Department of Community Development 46; dependence 5, 38, 39, 44, 59, 120, 146, 163; disaffiliation 4, 31, 86–8, 119, 124, 130, 131, 135, 138, 142, 157; domination 79, 80; elite 156, 164, 165, 167; employment 79–86, 124, 125; equality 59; evaluations 120, 131; executive committee 68; failure 111, 130; financial assistance 4, 48, 53, 57–60, 62–5, 68, 99, 102, 111, 112, 127, 128, 129, 150; financial dependency 123; financial mismanagement 69, 70, 118; financial problems 68–70, 72; guerillas 46; harambee 58, 64–6, 70, 115; history 45–88; homecraft center 50; humanitarianism 46, 59; income generating activities 60, 63, 111; income tax debt 69, 73, 202n143; indigenous agenda 135; inefficiency 69, 130, 142 201n136; Jeanes School 46, 47, 51, 52, 72; leadership 101, 157; linkages 112–13, 115, 120, 121, 144, 145, 148; linking pin 147; Lions Club, Kenya 60; literacy program 198n55; Maendeleo Handicraft Shop 57, 58; Maendeleo House 57; male domination 4, 63–88; membership 45, 46, 49, 50, 52, 55–7, 59, 110, 118; nation-building 117; national elected officers 77; national headquarters 4, 53, 67, 108–27, 130, 148; nepotism 69, 201n136; networks 127, 128, 130, 145; noblesse oblige 46; nursery school 56; oral tradition 198n55; patron 57, 87, 99; politicians 144; professionals 145; probe committee 69, 201n136; staff 101, 108, 120, 148; success 111, 130, 131; tea club 46, 160; technical assistance 4, 98, 99, 102 127, 129; transparency 130, 146; women's development body 70
MYWO national programs 115, 119; Leadership Development (LD) 58, 62, 87, 110, 112, 121, 124–6, 128, 137, 152; Maternal Child Health/Family Planning(MCH/FP) 58, 62, 110, 123, 124–6, 129, 131, 133, 135; Nutrition Program 58, 62, 70, 71, 87, 110, 121, 124, 126, 137; Special Energy Program-Jiko (SEP-Jiko) 58, 62, 110, 121, 124, 133, 137
MYWO 1989 elections 73–9; boycotts 76; election postponements 73, 74, 87; elected officials 77, 101, 111, 114, 134, 143, 144, 145–9, 148; KANU meddling 74–9; lobbyists 74–7, 85; protests 118, 120; queue voting 77, 203n144; voting schemes 74–7
magendo 93, 94, 157
Majengo 76
majimboism 97
Mali 25
Marsden, Peter 13
Marttaliitto 71, 87, 101, 127–31, 150–3, 162
Masale, Eliakim 69
Mathenge, Stanley 48
Matiba, Kenneth 60, 69, 71, 86
Mau Mau 46, 48–51; *see also* women
Manji, Julie 59

Maxwell, Robert 83
Mbogo, Jael 56
Mboya, Tom 53
McFadden, Patricia 160
Meru 49
Mitchell, J. C. 13
Mjomba, Joan 77, 101, 137, 143, 146
modernization theory *see* development
Mohammed, Fatuma 82
Moi, Daniel arap 90, 91, 117; Nyayo philosophy 66
Moi Day 74
Molo 97
Moore, Winifred 48
Moser, Caroline 34
Mother's Hunger Strike *see* Kenyan government opposition
Mugo, Beth 95, 142
Mugulla, Joseph 92
Muite, Paul 95
Mulford, Charles 9, 158
Mulindi, Gladys 69
multipartyism 80, 82, 83, 95–7, 123, 125, 142, 146, 160; Democratic Movement (DEMO) 96; Democratic Party (DP) 86, 96; Democratization 96; Forum for the Restoration of Democracy (FORD) 86, 96; Islamic Party of Kenya (IPK) 96; Social Democratic Party (SDP) 96
Muranga 63, 81
Musoke, Harriet 47
Mutisya, Mulu 74
Mutunga, Willy 95
Mwamodo, Mary 69
Mwangi, Jacob 26
Mwenda, Elizabeth 56
Mwenda, Kitili 57
Mwenda, Nyiva 57
Myrdal, Gunnar 31

Nairobi City Commission 23
Nairobi International Show 74
Nairobi Law Monthly 95
Nairobi Trade Bank 91, 92
Nakuru 49
Nandi *see also* ethnicity 97
Narok 74
Nasimiyu, Ruth 186n10
Nassir, Shariff 76, 85
National Christian Council of Kenya (NCCK) 67, 95
National Council of Women of Kenya (NCWK) 65, 67, 68, 83, 84, 144
Ndegwa, Stephen 15, 28
Ndetei, Agnes 86
Ndilinge, Tony 74
neocolonialism 5
Netherlands 82
Ngai Murunya Ward 75
Njenga, Jonathan 56
Nkrumah, Kwame 53
non-governmental organization (NGO): autonomy 5, 6, 21, 24, 45, 53, 78, 99, 130, 139, 140, 153, 166, 167; branches 20, 21; Coordination Act 1990 28, 125, 131, 142; cultural imperialism 20; definition 21, 149, 150, 152, harassment 19; European 17; Northern 20–3, 27, 152, 155; Southern 17, 19, 20–23, 27, 152, 155; Task Force for Eastern and Southern Africa 142
non-governmental development organization (NGDOs) 21, 23
North, global 12, 155, 156, 160, 161, 163, 166, 167, 188n20
North Eastern Province 113
North Korea 97
Norweigan Agency for International Development (NORAD) 62, 87, 114, 124, 136, 153, 154
Nyangoro, Julius 14–15
Nyanza 49, 50
Nyayo philosophy 66, 86 *see also* Kenyan goverment; Moi
Nyeri 49, 75, 85
Nyoni, Sithembiso 161
Nzomo, Maria 35, 37, 55, 85, 147, 167

Odinga, Oginga 86, 90
Ogot, Grace 82, 86
Ojiambo, Julia 73, 74, 77
Ombaka, Ooka 94
Ombara, Gladys 50
Ongata Rongai Ward 75
Onsando, Wilkista 63, 77, 80–4, 87, 101, 107, 109, 114, 112, 116, 122, 128, 132, 145, 146, 132, 141
Organization for African Unity (OAU) 92
organizational theory: relational evironment 13–15
Orora, John 64

Otete, Francisca 70, 73
Otieno, S. M. 82
Otieno, Wambui 82
Ouko, Robert 92–4
Oxford Committee for Famine Relief (OXFAM) 22, 60
Oyugi, Hezekiah 91, 94

Padron, Mario 23
pan-Africanism 53
Pappi, Franz 14
Paris 96
Parpart, Jane 32, 33, 36, 167
Parsons, Talcott 41
partnerships *see* development partnerships
Pathfinder International 101, 123, 127, 128
patriarchy: global 35, 41; Kenyan 35; *see also* gender; Kenya Africa National Union; Kenyan goverment
patronage 18
political culture 28
political economy 15, 39, 31
Penwill, Mrs 49
power *see* development partnerships; interorganizational relations; RDM
Pradervand, Pierre 4
Presley, Cora 46
Pringle, Rosemary 36

Quarry Ward 75

Ramsey, Karen 36
research strategy 104–7; archives 6, 89, 47, 48, 103; case study approach 4, 105–7; data 47, 48, 99–106; deductive logic 105; dependent variables 98, 172, 174; field observations 103, 104; fieldwork 98–104; hypotheses 89, 98, 105, 169–72; independent variables 98, 172–4; inductive logic 105; intervening variables 98, 172–4; interviewees anonymity xii, 8; interviewees feedback xii, 8; methodology 98–107; multi-method strategy 105–6; normative school 104, 195; ontology 104; paradigms 104; political climate 89–98; positivist school 104, 105; postpositivist school 104, 105; qualitative methods 104–5; quantitative methods 105; snowball method 106, 107; study instrumentation 100–5, 89; sub-hypotheses 170–1; suspicion of foreign researchers 89–98; survey instrument 100, 175–8
Resource Dependency Model (RDM) 38–44, 158–9, 167; control 41; dependence 154, 167; exchange 39–40; power 38–44; theoretical framework 38–44;
Rift Valley Province 97
Rodney, Walter 31
Rogers, Barbara 31, 37
Rural Development Fund 112

Saitoti, George 81, 91, 93, 96, 97,
Sala, John 75
Salvation Army 22
Saunders, Paul 9, 12, 13
Savage, Mike 35
Sawyer, W. H. 50
Scheme Six Ward 75
Schneider, Bertrand 161
Scott, Catherine 32–3, 147
Section 2(a) of Kenyan Constitution 95, 97
self-reliance 5, 61
Senegal 25
Shepherd, Nancy 46, 52, 59
Shimanzi 76
Shikaku, Martin 86
Shitakha, Theresa 68–71, 84, 99, 107, 118
Smillie, Ian 10, 19
Smith, Brian 15
Smith, William E. 17
Social Democratic Party (SDP) 96
Somaia, Ketan 91
South, global 12, 156, 160, 161, 163, 166, 167, 168, 188n20
Speigel, Hans 64
The Standard 69, 103
Staudt, Kathleen 36, 37, 167
Steel, David 26
Stoltenberg, John 34–5
structural adjustment programs (SAPs) 92; economic liberalization 99, 146
Sumba, Zubeda 76
Sweden 82
Switzerland 82

Taita Taveta 50
Tambach 49
Tandon, Yash 27, 154
Tanzania 61
technical assistance *see also* foreign donors; Kenyan government; Maendeleo Ya Wanawake
Tendler, Judith 18
Thiele, Graham 15
Thoolen, Ben E. 18
Transnational corporation (TNC) 23
transnational relations 13
Twose, Nigel 17, 19, 20

Uhuru Park 83 *see also* Maathai, Wangari
underdevelopment *see* development
Undugu Society 28
United Kingdom 82, 90
United Nations (UN) 150
United Nations Decade for Women 60, 65, 67, 69, 202n143 *see also* women
United Nations Conference on Trade and Development (UNCTD) 12
United Nations Development Programme (UNDP) 82
United Nations International Children's Emergency Fund (UNICEF) 50, 53
United States Agency for International Development (USAID) 22, 62, 81, 101, 110, 123, 124, 126–8, 150, 131
United States Peace Corps 22, 101, 127–9, 131, 133, 149, 150, 151–4
Uphoff, Norman 28, 166

values 55, 166
Van de Ven, Andrew 14
Van der Heijden, Hendrik 26
Vincent, Ferdinand 26
Viva Magazine 73
voluntary development organization (VDO) 16

Waiganjo, Fred 77
Wallerstein, Immanuel 31
Wamwere, Koigi wa 87, 114, 136
Wandega, Ester 77
Wanjiru, Lydiah 75
Wanyeki, Grace 61
Waruhiu, Rose 76
Weekly Review 76, 95
Wellard, Kate 15
White Highlands 47
White, Paul 39, 41
Williams, Louise 49
Wipper, Audrey 4, 46, 99
Witz, Anne 35
women: childcare 35; Decade for Women 60, 65, 67, 69, 202n143; dependence 34, 38, 40, 44, 73; development assistance 29, 30, 38, 39, 60–3, 68, 86–8; economic conditions 3, 32; education 198n55; elections *see also* MYWO 1989 elections; entrepreneurship 141; equality 1, 34, 59; exit 38; grassroots 32, 88, 108, 114, 124, 125, 144, 157, 160, 161; groups 45–63, 98, 100–1, 108–14, 127, 132, 133, 139, 141, 165, 166, 168; guerillas 46; leaders 52, 54–63, 77–9, 82–5, 86, 205n181; liberation 37, 59; mabati women's groups 57; male domination 37, 124; male members 110; Mau Mau 47–8; merry-go-round 141; national identity cards 73; politicians 32; power 46, 107, 113; protesters 75–6; rights 59; secret society 48; self-help services 53, 56, 57; voters 73–9, 67; younger 141, 142
Women and Development (WAD) 33, 166
Women's Bureau *see* Kenyan Government
Women in Development (WID) 32, 33, 65, 66, 141, 144, 166
World Bank 82, 92, 127, 128, 133, 140, 150, 151, 154, 158, 163
world systems theory *see* development

Yudelman, Sally 37

Zimbabwe 17–19, 91, 164